ESSAY   T0166717   M

ALBERT EINSTEIN

Essays in Humanism

INTEGRATED MEDIA

NEW YORK

Cover design by Milan Bozic
Text design and composition by India Amos

ISBN 978-1-4532-0463-4

Philosophical Library
119 W. 40th Street
New York, NY 10018
www.philosophicallibrary.com

Distributed by Open Road Integrated Media, Inc.
345 Hudson Street
New York, NY 10014
www.openroadmedia.com

Contents

ESSAYS IN HUMANISM

1

Why Socialism?

Is it advisable for one who is not an expert on economic and social issues to express views on the subject of socialism? I believe for a number of reasons that it is.

Let us first consider the question from the point of view of scientific knowledge. It might appear that there are no essential methodological differences between astronomy and economics: scientists in both fields attempt to discover laws of general acceptability for a circumscribed group of phenomena in order to make the interconnection of these phenomena as clearly understandable as possible. But in reality such methodological differences do exist. The discovery of general laws in the field of economics is made difficult by the circumstance that observed economic phenomena are often affected by many factors which are very hard to evaluate separately. In addition, the experience which has accumulated since the beginning of the so-called civilized period of human history has—as is well known—been largely influenced and limited by causes which are by no means exclusively economic in nature. For example, most of the major states of history owed their existence to conquest. The conquering peoples established themselves, legally and economically, as the privileged class of the conquered country. They seized for themselves a monopoly of the land ownership and appointed a priesthood from among their own ranks. The priests, in control of education, made the class division of society into a permanent institution and created a system of values by which the people were thenceforth, to a large extent unconsciously, guided in their social behavior.

But historic tradition is, so to speak, of yesterday; nowhere have we really overcome what Thorstein Veblen called "the predatory phase" of human development. The observable economic facts belong to that phase and even such laws as we can derive from them are not applicable to other phases. Since the real purpose of socialism is precisely to overcome and advance beyond the predatory phase of human development, economic science in its present state can throw little light on the socialist society of the future.

Second, socialism is directed towards a social-ethical end. Science, however, cannot create ends and, even less, instill them in human beings; science, at most, can supply the means by which to attain certain ends. But the ends themselves are conceived by personalities with lofty ethical ideals and—if these ends are not stillborn, but vital and vigorous—are adopted and carried forward by those many human beings who, half unconsciously, determine the slow evolution of society.

For these reasons, we should be on our guard not to overestimate science and scientific methods when it is a question of human problems; and we should not assume that experts are the only ones who have a right to express themselves on questions affecting the organization of society.

Innumerable voices have been asserting for some time now that human society is passing through a crisis, that its stability has been gravely shattered. It is characteristic of such a situation that individuals feel indifferent or even hostile toward the group, small or large, to which they belong. In order to illustrate my meaning, let me record here a personal experience. I recently discussed with an intelligent and well-disposed man the threat of another war, which in my opinion would seriously endanger the existence of mankind, and I remarked that only a supranational organization would offer protection from that danger. Thereupon my visitor, very calmly and coolly, said to me: "Why are you so deeply opposed to the disappearance of the human race?"

I am sure that as little as a century ago no one would have so lightly made a statement of this kind. It is the statement of a man who has striven in vain to attain an equilibrium within himself and has more or less lost hope of succeeding. It is the expression of a painful solitude and isolation from which so many people are suffering in these days. What is the cause? Is there a way out?

It is easy to raise such questions, but difficult to answer them with any degree of assurance. I must try, however, as best I can, although I am very conscious of the fact that our feelings and strivings are often contradictory and obscure and that they cannot be expressed in easy and simple formulas.

Man is, at one and the same time, a solitary being and a social being. As a solitary being, he attempts to protect his own existence and that of those who are closest to him, to satisfy his personal desires, and to develop his innate abilities. As a social being, he seeks to gain the recognition and affection of his fellow human beings, to share in their pleasures, to comfort them in their sorrows, and to improve their conditions of life. Only the existence of these varied, frequently conflicting, strivings accounts for the special character of a man, and their specific combination determines the extent to which an individual can achieve an inner equilibrium and can contribute to the well-being of society. It is quite possible that the relative strength of these two drives is, in the main, fixed by inheritance. But the personality that finally emerges is largely formed by the environment in which a man happens to find himself during his development, by the structure of the society in which he grows up, by the tradition of that society, and by its appraisal of particular types of behavior. The abstract concept "society" means to the individual human being the sum total of his direct and indirect relations to his contemporaries and to all the people of earlier generations. The individual is able to think, feel, strive, and work by himself; but he depends so much upon society—in his physical, intellectual, and emotional existence—that it is impossible to think of him, or to understand him, outside the framework of society. It is "society" which provides man with food, clothing, a home, the tools of work, language, the forms of thought, and most of the content of thought; his life is made possible through the labor and the accomplishments of the many millions past and present who are all hidden behind the small word "society."

It is evident, therefore, that the dependence of the individual upon society is a fact of nature which cannot be abolished—just as in the case of ants and bees. However, while the whole life process of ants and bees is fixed down to the smallest detail by rigid, hereditary instincts, the social pattern and interrelationships of human beings are very variable and susceptible to change. Memory, the capacity to make new combinations, the gift of oral communication have made possible developments among human beings which are not dictated by biological necessities. Such developments manifest themselves in traditions, institutions, and organizations; in literature; in scientific and engineering accomplishments; in works of art. This explains how it happens that, in a certain sense, man can influence his life through his own conduct, and that in this process conscious thinking and wanting can play a part.

Man acquires at birth, through heredity, a biological constitution which we must consider fixed and unalterable, including the natural urges which are characteristic of the human species. In addition, during his lifetime, he acquires a cultural constitution which he adopts from society through communication and through many other types of influences. It is this cultural constitution which, with the passage of time, is subject to change and which determines to a very large extent the relationship between the individual and society. Modern anthropology has taught us, through comparative investigation of so-called primitive cultures, that the social behavior of human beings may differ greatly, depending upon prevailing cultural patterns and the types of organization which predominate in society. It is on this that those who are striving to improve the lot of man may ground their hopes: human beings are not condemned, because of their biological constitution, to annihilate each other or to be at the mercy of a cruel, self-inflicted fate.

If we ask ourselves how the structure of society and the cultural attitude of man should be changed in order to make human life as satisfying as possible, we should constantly be conscious of the fact that there are certain conditions which we are unable to modify. As mentioned before, the biological nature of man is, for all practical purposes, not subject to change. Furthermore, technological and demographic developments of the last few centuries have created conditions which are here to stay. In relatively densely settled populations with the goods which are indispensable to their continued existence, an extreme division of labor and a highly-centralized productive apparatus are absolutely

necessary. The time—which, looking back, seems so idyllic—
is gone forever when individuals or relatively small groups
could be completely self-sufficient. It is only a slight exaggeration to say that mankind constitutes even now a planetary
community of production and consumption.

I have now reached the point where I may indicate briefly
what to me constitutes the essence of the crisis of our time.
It concerns the relationship of the individual to society. The
individual has become more conscious than ever of his
dependence upon society. But he does not experience this
dependence as a positive asset, as an organic tie, as a pro-
tective force, but rather as a threat to his natural rights, or
even to his economic existence. Moreover, his position in
society is such that the egotistical drives of his make-up are
constantly being accentuated, while his social drives, which
are by nature weaker, progressively deteriorate. All human
beings, whatever their position in society, are suffering from
this process of deterioration. Unknowingly prisoners of their
own egotism, they feel insecure, lonely, and deprived of the
naive, simple, and unsophisticated enjoyment of life. Man can
find meaning in life, short and perilous as it is, only through
devoting himself to society.

The economic anarchy of capitalist society as it exists
today is, in my opinion, the real source of the evil. We see
before us a huge community of producers the members of
which are unceasingly striving to deprive each other of the
fruits of their collective labor—not by force, but on the whole
in faithful compliance with legally established rules. In this
respect, it is important to realize that the means of production—that is to say, the entire productive capacity that is
needed for producing consumer goods as well as additional

capital goods—may legally be, and for the most part are, the private property of individuals.

For the sake of simplicity, in the discussion that follows I shall call "workers" all those who do not share in the ownership of the means of production—although this does not quite correspond to the customary use of the term. The owner of the means of production is in a position to purchase the labor power of the worker. By using the means of production, the worker produces new goods which become the property of the capitalist. The essential point about this process is the relation between what the worker produces and what he is paid, both measured in terms of real value. Insofar as the labor contract is "free," what the worker receives is determined not by the real value of the goods he produces, but by his minimum needs and by the capitalists' requirements for labor power in relation to the number of workers competing for jobs. It is important to understand that even in theory the payment of the worker is not determined by the value of his product.

Private capital tends to become concentrated in few hands, partly because of competition among the capitalists, and partly because technological development and the increasing division of labor encourage the formation of larger units of production at the expense of the smaller ones. The result of these developments is an oligarchy of private capital the enormous power of which cannot be effectively checked even by a democratically organized political society. This is true since the members of legislative bodies are selected by political parties, largely financed or otherwise influenced by private capitalists who, for all practical purposes, separate the electorate from the legislature. The consequence is that the representatives of the people do not in fact sufficiently protect the interests of the underprivileged sections of the population. Moreover, under existing conditions, private capitalists inevitably control, directly or indirectly, the main sources of information (press, radio, education). It is thus extremely difficult, and indeed in most cases quite impossible, for the individual citizen to come to objective conclusions and to make intelligent use of his political rights.

The situation prevailing in an economy based on the private ownership of capital is thus characterized by two main principles: first, means of production (capital) are privately owned and the owners dispose of them as they see fit; second, the labor contract is free. Of course, there is no such thing as a pure capitalist society in this sense. In particular, it should be noted that the workers, through long and bitter political struggles, have succeeded in securing a somewhat improved form of the "free labor contract" for certain categories of workers. But taken as a whole, the present day economy does not differ much from "pure" capitalism.

Production is carried on for profit, not for use. There is no provision that all those able and willing to work will always be in a position to find employment; an "army of unemployed" almost always exists. The worker is constantly in fear of losing his job. Since unemployed and poorly paid workers do not provide a profitable market, the production of consumers' goods is restricted, and great hardship is the consequence. Technological progress frequently results in more unemployment rather than in an easing of the burden of work for all. The profit motive, in conjunction with competition among capitalists, is responsible for an instability in the accumulation and utilization of capital which leads to increasingly severe depressions. Unlimited competition leads to a huge waste of labor, and to that crippling of the social consciousness of individuals which I mentioned before.

This crippling of individuals I consider the worst evil of capitalism. Our whole educational system suffers from this evil. An exaggerated competitive attitude is inculcated into the student, who is trained to worship acquisitive success as a preparation for his future career.

I am convinced there is only one way to eliminate these grave evils, namely through the establishment of a socialist economy, accompanied by an educational system which would be oriented toward social goals. In such an economy, the means of production are owned by society itself and are utilized in a planned fashion. A planned economy, which adjusts production to the needs of the community, would distribute the work to be done among all those able to work and would guarantee a livelihood to every man, woman, and child. The education of the individual, in addition to promoting his own innate abilities, would attempt to develop in him a sense of responsibility for his fellow men in place of the glorification of power and success in our present society.

Nevertheless, it is necessary to remember that a planned economy is not yet socialism. A planned economy as such may be accompanied by the complete enslavement of the individual. The achievement of socialism requires the solution of some extremely difficult socio-political problems: how is it possible, in view of the far-reaching centralization of political and economic power, to prevent bureaucracy from becoming all-powerful and overweening? How can the rights of the individual be protected and therewith a democratic counterweight to the power of bureaucracy be assured?

2

The Negro Question

I AM WRITING as one who has lived among you in America only a little more than ten years. And I am writing seriously and warningly. Many readers may ask: "What right has he to speak out about things which concern us alone, and which no newcomer should touch?"

I do not think such a standpoint is justified. One who has grown up in an environment takes much for granted. On the other hand, one who has come to this country as a mature person may have a keen eye for everything peculiar and characteristic. I believe he should speak out freely on what he sees and feels, for by so doing he may perhaps prove himself useful.

What soon makes the new arrival devoted to this country is the democratic trait among the people. I am not thinking here so much of the democratic political constitution of this country, however highly it must be praised. I am thinking of the relationship between individual people and of the attitude they maintain toward one another.

In the United States everyone feels assured of his worth as an individual. No one humbles himself before another person or class. Even the great difference in wealth, the superior power of a few, cannot undermine this healthy self-confidence and natural respect for the dignity of one's fellow-man.

There is, however, a somber point in the social outlook

of Americans. Their sense of equality and human dignity is mainly limited to men of white skins. Even among these there are prejudices of which I as a Jew am clearly conscious; but they are unimportant in comparison with the attitude of the "Whites" toward their fellow-citizens of darker complexion, particularly toward Negroes. The more I feel an American, the more this situation pains me. I can escape the feeling of complicity in it only by speaking out.

Many a sincere person will answer me: "Our attitude towards Negroes is the result of unfavorable experiences which we have had by living side by side with Negroes in this country. They are not our equals in intelligence, sense of responsibility, reliability."

I am firmly convinced that whoever believes this suffers from a fatal misconception. Your ancestors dragged these black people from their homes by force; and in the white man's quest for wealth and an easy life they have been ruthlessly suppressed and exploited, degraded into slavery. The modern prejudice against Negroes is the result of the desire to maintain this unworthy condition.

The ancient Greeks also had slaves. They were not Negroes but white men who had been taken captive in war. There could be no talk of racial differences. And yet Aristotle, one of the great Greek philosophers, declared slaves inferior beings who were justly subdued and deprived of their liberty. It is clear that he was enmeshed in a traditional prejudice from which, despite his extraordinary intellect, he could not free himself.

A large part of our attitude toward things is conditioned by opinions and emotions which we unconsciously absorb as children from our environment. In other words, it is tradition—besides inherited aptitudes and qualities—which makes us what we are. We but rarely reflect how relatively small as compared with the powerful influence of tradition is the influence of our conscious thought upon our conduct and convictions.

It would be foolish to despise tradition. But with our growing self-consciousness and increasing intelligence we must begin to control tradition and assume a critical attitude toward it, if human relations are ever to change for the better. We must try to recognize what in our accepted tradition is damaging to our fate and dignity—and shape our lives accordingly.

I believe that whoever tries to think things through honestly will soon recognize how unworthy and even fatal is the traditional bias against Negroes.

What, however, can the man of good will do to combat this deeply rooted prejudice? He must have the courage to set an example by word and deed, and must watch lest his children become influenced by this racial bias.

I do not believe there is a way in which this deeply entrenched evil can be quickly healed. But until this goal is reached there is no greater satisfaction for a just and well-meaning person than the knowledge that he has devoted his best energies to the service of the good cause.

3

Science and Society

————————

THERE ARE TWO WAYS in which science affects human affairs. The first is familiar to everyone: Directly, and to an even greater extent indirectly, science produces aids that have completely transformed human existence. The second way is educational in character—it works on the mind. Although it may appear less obvious to cursory examination, it is no less incisive than the first.

The most conspicuous practical effect of science is that it makes possible the contriving of things that enrich life, though they complicate it at the same time—inventions such as the steam engine, the railway, electric power and light, the telegraph, radio, automobile, airplane, dynamite, etc. To these must be added the life-preserving achievements of biology and medicine, especially the production of pain relievers and preservative methods of storing food. The greatest practical benefit which all these inventions confer on man I see in the fact that they liberate him from the excessive muscular drudgery that was once indispensable for the preservation of bare existence. Insofar as we may at all claim that slavery has been abolished today, we owe its abolition to the practical consequences of science.

On the other hand, technology—or applied science—has confronted mankind with problems of profound gravity. The very survival of mankind depends on a satisfactory solution of these problems. It is a matter of creating the kind of social institutions and traditions without which the new tools must inevitably bring disaster of the worst kind.

Mechanical means of production in an unorganized economy have had the result that a substantial proportion of mankind is no longer needed for the production of goods and is thus excluded from the process of economic circulation. The immediate consequences are the weakening of purchasing power and the devaluation of labor because of excessive competition, and these give rise, at ever shortening intervals, to a grave paralysis in the production of goods. Ownership of the means of production, on the other hand, carries a power to which the traditional safeguards of our political institutions are unequal. Mankind is caught up in a struggle for adaptation to these new conditions—a struggle that may bring true liberation, if our generation shows itself equal to the task.

Technology has also shortened distances and created new and extraordinarily effective means of destruction which, in the hands of nations claiming unrestricted freedom of action, become threats to the security and very survival of mankind. This situation requires a single judicial and executive power for the entire planet, and the creation of such a central authority is desperately opposed by national traditions. Here too we are in the midst of a struggle whose issue will decide the fate of all of us.

Means of communication, finally—reproduction processes for the printed word, and the radio—when combined with modern weapons, have made it possible to place body and soul under bondage to a central authority—and here is a third source of danger to mankind. Modern tyrannies and their destructive effects show plainly how far we are from exploiting these achievements organizationally for the benefit of mankind. Here too circumstances require an international solution, with the psychological foundation for such a solution not yet laid.

Let us now turn to the intellectual effects that proceed from science. In prescientific times it was not possible by means of thought alone to attain results that all mankind could have accepted as certain and necessary. Still less was there a conviction that all that happens in nature is subject to inexorable laws. The fragmentary character of natural law, as seen by the primitive observer, was such as to foster a belief in ghosts and spirits. Hence even today primitive man lives in constant fear that supernatural and arbitrary forces will intervene in his destiny.

It stands to the everlasting credit of science that by acting on the human mind it has overcome man's insecurity before himself and before nature. In creating elementary mathematics the Greeks for the first time wrought a system of thought whose conclusions no one could escape. The scientists of the Renaissance then devised the combination of systematic experiment with mathematical method. This union made possible such precision in the formulation of natural laws and such certainty in checking them by experience that as a result there was no longer room for basic differences of opinion in natural science. Since that time each generation has built up the heritage of knowledge and understanding, without the slightest danger of a crisis that might jeopardize the whole structure.

The general public may be able to follow the details of scientific research to only a modest degree; but it can register at least one great and important gain: confidence that human thought is dependable and natural law universal.

4

Towards a World Government

A CONVERSATION I HAD with three students of the University of Chicago has made a strong impression on me. It showed me that a sense of responsibility and initiative is at work in the young generation of this country. These students are aware of the fact that the destiny of the new generation, will be decided in these few years. They are determined to influence the pace of events within the framework of their possibilities.

What is the situation? The development of technology and of the implements of war has brought about something akin to a shrinking of our planet. Economic interlinking has made the destinies of nations interdependent to a degree far greater than in previous years. The available weapons of destruction are of a kind such that no place on earth is safeguarded against sudden total destruction. The only hope for protection lies in the securing of peace in a supranational way. A world government must be created which is able to solve conflicts between nations by judicial decision. This government must be based on a clearcut constitution which is approved by the governments and the nations and which gives it the sole disposition of offensive weapons. A person or a nation can be considered peace loving only if it is ready to cede its military force to the international authorities and to renounce every attempt or even the means, of achieving its interests abroad by the use of force.

It is apparent that the development of political relations in the year which has elapsed since the conclusion of the second world war, has brought us in no way nearer to the achievement of this goal. The u. n. as it stands today has neither the military force nor the legal basis to bring about a state of international security. Nor does it take account of the actual distribution of power. Real power is at present in the hands of few. It is no exaggeration to say that the solution of the real problem is linked solely to an agreement on a grand scale between this country and Russia. For, if such an agreement would be achieved then these two powers alone would be able to cause the other nations to give up their sovereignty to the degree necessary for the establishment of military security for all.

Now many will say that fundamental agreement with Russia is impossible under the present circumstances. Such a statement would be justified if the United States had made a serious attempt in this direction during the past year. I find, however, that the opposite has happened. There was no need to accept fascist Argentina into the u. n. against Russia's opposition. There was no need to manufacture new atomic bombs without letup and to appropriate twelve billion dollars for defense in a year in which no military threat was to be expected for the nearest future. Nor was it necessary to delay the proposed measures against Franco-Spain. It is senseless to recount here the details which all show that nothing has been done in order to alleviate Russia's distrust, a distrust which can very well be understood in the light of the events of the last decades and to whose origin we have contributed no little.

A permanent peace cannot be prepared by threats but only by the honest attempt to create mutual trust. One should think that the wish to create a decent form of life on this planet and to avert the danger of unspeakable destruction would tame the passions of responsible men. You cannot rely on that, my young friends. May you succeed in activating the young generation in this sense, so that it will strive for a policy of peace on a grand scale. Thus you can not only defend yourself successfully but you can serve your country and your descendants in a degree as was not given to any previous generation.

5

The Way Out

THE CONSTRUCTION OF THE atom bomb has brought
about the effect that all the people living in cities are threatened, everywhere and constantly, with sudden destruction.
There is no doubt that this condition has to be abolished
if man is to prove himself worthy, at least to some extent,
of the self-chosen name of homo sapiens. However, there
still exist widely divergent opinions concerning the degree
to which traditional social and political forms, historically
developed, will have to be sacrificed in order to achieve the
desired security.

After the First World War, we were confronted with a
paradoxical situation regarding the solution of international
conflicts. An international court of justice had been established for a peaceful solution of these conflicts on the basis
of international law. Furthermore, a political instrument for
securing peace by means of international negotiation in a
sort of world parliament had been created in the form of
the League of Nations. The nations united in the League had
further outlawed as criminal the method of solving conflicts
by means of war.

Thus the nations were imbued with an illusion of security
that led inevitably to bitter disappointment. For the best
court of justice is meaningless unless it is backed by the
authority and power to execute its decisions, and exactly
the same thing is true of a world parliament. An individual

state with sufficient military and economic power can easily resort to violence and voluntarily destroy the entire structure of supranational security built on nothing but words and documents. Moral authority alone is an inadequate means of securing the peace.

The United Nations Organization is now in the process of being tested. It may eventually emerge as the agency of "security without illusion" that we so badly need. But it has not as yet gone beyond the area of moral authority as, in my opinion, it must.

Our situation is rendered more acute by other circumstances, only two of which will be presented here. So long as the individual state, despite its official condemnation of war, has to consider the possibility of engaging in war, it must influence and educate its citizens—and its youth in particular—in such a way that they can easily be converted into efficient soldiers in the event of war. Therefore it is compelled not only to cultivate a technical-military training and type of thinking but also to implant a spirit of national vanity in its people in order to secure their inner readiness for the outbreak of war. Of course, this kind of education counteracts all endeavors to establish moral authority for any supranational security organization.

The danger of war in our time is further heightened by another technical factor. Modern weapons, in particular the atom bomb, have led to a considerable advantage in the means of offense or attack over those of defense. And this could well bring about the result that even responsible statesmen might find themselves compelled to wage a preventive war.

In view of these evident facts there is, in my opinion, only one way out.

It is necessary that conditions be established that guarantee the individual state the right to solve its conflicts with other states on a legal basis and under international jurisdiction.

It is necessary that the individual state be prevented from making war by a supranational organization supported by a military power that is exclusively under its control.

Only when these two conditions have been fully met can we have some assurance that we shall not vanish into the atmosphere, dissolved into atoms, one of these days.

From the viewpoint of the political mentality prevailing at present, it may seem illusory, even fantastic, to hope for the realization of such conditions within a period of a few years. Yet their realization cannot wait for a gradual historical development to take its course. For, so long as we do not achieve supranational military security, the above-mentioned factors can always and forcibly lead us into war. Even more than the will for power, the fear of sudden attack will prove to be disastrous for us if we do not openly and decisively meet the problem of depriving national spheres of power of their military strength, turning such power over to a supranational authority.

With due consideration for the difficulties involved in this task, I have no doubt about one point. We shall be able to solve the problem when it will be clearly evident to all that there is no other, no cheaper way out of the present situation.

Now I feel it my obligation to say something about the individual steps which might lead to a solution of the security problem.

1. Mutual inspection by the leading military powers of methods and installations used for the production of

offensive weapons, combined with an interchange of pertinent technical and scientific discoveries, would diminish fear and distrust, at least for the time being. In the breathing spell thus provided we would have to prepare more thorough measures. For this preliminary step should be taken with conscious awareness that the ultimate goal is the denationalization of military power altogether.

This first step is necessary to make any successive moves possible. However, we should be wary of believing that its execution would immediately result in security. There still would remain the possibility of an armament race with regard to a possible future war, and there always exists the temptation to resort once more, by "underground" methods, to the military secret, that is, keeping secret the knowledge about methods and means of and actual preparations for warfare. Real security is tied to the denationalization of military power.

2. This denationalization can be prepared through a steadily increasing interchange of military and scientific-technical personnel among the armies of the different nations. The interchange should follow a carefully elaborated plan, aimed at converting the national armies systematically into a supranational military force. A national army, one might say, is the last place where national feeling may be expected to weaken. Even so, the nationalism can be progressively immunized at a rate proportionate at least to the building of the supranational army; and the whole process can be facilitated by integrating it with the recruiting and training of the latter. The process of interchanging personnel would further lessen the danger of

surprise attacks and in itself would lay the psychological foundation for internationalization of military resources.

Simultaneously the strongest military powers could draft the working papers for a supranational security organization and for an arbitration committee, as well as the legal basis for, and the precise stipulation of, obligations, competencies, and restrictions of the latter with respect to the individual nations. They could further decide upon the terms of election for establishing and maintaining these bodies.

When an agreement on these points shall have been reached, a guarantee against wars of world-wide dimensions can be assured.

3. The above-named bodies can now begin to function. The vestiges of national armies can then be either disbanded or placed under the high command of the supranational authority.

4. After the cooperation of the nations of highest military importance has been secured, the attempt should be made to incorporate, if possible, all nations into the supranational organization, provided that it is their voluntary decision to join.

This outline may perhaps create the impression that the presently prevailing military powers are to be assigned too dominant a role. I have tried, however, to present the problem with a view to a sufficiently swift realization that will allow us to avoid difficulties greater than those already inherent in the nature of such a task. It may be simpler, of course, to reach preliminary agreement among the strongest military powers than among all nations, big and small, for a body of

representatives of all nations is a hopelessly clumsy instrument for the speedy achievement of even preliminary results.
Even so, the task confronting us requires of all concerned the
utmost sagacity and tolerance, which can be achieved only
through awareness of the harsh necessity we have to face.

6

On Receiving the One World Award

I AM GREATLY TOUCHED by the signal honor which you have wished to confer upon me. In the course of my long life I have received from my fellow-men far more recognition than I deserve, and I confess that my sense of shame has always outweighed my pleasure therein. But never, on any previous occasion, has the pain so far outweighed the pleasure as now. For all of us who are concerned for peace and the triumph of reason and justice must today be keenly aware how small an influence reason and honest good will exert upon events in the political field. But however that may be, and whatever fate may have in store for us, yet we may rest assured that without the tireless efforts of those who are concerned with the welfare of humanity as a whole, the lot of mankind would be still worse than in fact it even now is.

In this time of decisions so heavy with fate what we must say to our fellow-citizens seems above all to be this: where belief in the omnipotence of physical force gets the upper hand in political life this force takes on a life of its own, and proves stronger than the men who think to use force as a tool. The proposed militarization of the nation not only immediately threatens us with war; it will also slowly but surely destroy the democratic spirit and the dignity of the individual in our land. The assertion that events abroad force us to arm is wrong, we must combat it with all our strength. Actually, our own rearmament, through the reaction of other nations to it, will bring about that very situation on which its advocates seek to base their proposals.

There is only one path to peace and security: the path of supra-national organization. One-sided armament on a national basis only heightens the general uncertainty and confusion without being an effective protection.

7

Science and Civilizations

IT IS IN TIMES of economic distress such as we experience everywhere today, one sees very clearly the strength of the moral forces that live in a people. Let us hope that a historian delivering judgment in some future period when Europe is politically and economically united, will be able to say that in our days the liberty and honour of this Continent was saved by its Western nations, which stood fast in hard times against the temptations of hatred and oppression; and that Western Europe defended successfully the liberty of the individual which has brought us every advance of knowledge and invention—liberty without which life to a self-respecting man is not worth living.

It cannot be my task today to act as judge of the conduct of a nation which for many years has considered me as her own; perhaps it is an idle task to judge in times when action counts.

Today, the questions which concern us are: how can we save mankind and its spiritual acquisitions of which we are the heirs? How can one save Europe from a new disaster?

It cannot be doubted that the world crisis and the suffering and privations of the people resulting from the crisis are in some measure responsible for the dangerous upheavals of which we are the witness. In such periods discontent breeds hatred, and hatred leads to acts of violence and revolution, and often even to war. Thus distress and evil produce new

distress and new evil. Again the leading statesmen are bur-
dened with tremendous responsibilities just the same as
twenty years ago. May they succeed through timely agreement to establish a condition of unity and clarity of international obligations in Europe, so that for every State a warlike
adventure must appear as utterly hopeless. But the work of
statesmen can succeed only if they are backed by the serious
and determined will of the people.

We are concerned not merely with the technical problem of securing and maintaining peace, but also with the
important task of education and enlightenment. If we want to
resist the powers which threaten to suppress intellectual and
individual freedom we must keep clearly before us what is at
stake, and what we owe to that freedom which our ancestors
have won for us after hard struggles.

Without such freedom there would have been no Shake-
speare, no Goethe, no Newton, no Faraday, no Pasteur and
no Lister. There would be no comfortable houses for the mass
of the people, no railway, no wireless, no protection against
epidemics, no cheap books, no culture and no enjoyment of
art for all. There would be no machines to relieve the people
from the arduous labor needed for the production of the
essential necessities of life. Most people would lead a dull
life of slavery just as under the ancient despotisms of Asia.
It is only men who are free, who create the inventions and
intellectual works which to us moderns make life worth while.

Without doubt the present economic difficulties will
eventually bring us to the point where the balance between
supply of labor and demand of labor, between production and
consumption, will be enforced by law. But even this problem
we shall solve as free men and we shall not allow ourselves for

its sake to be driven into a slavery, which ultimately would bring with it stagnation of every healthy development.

In this connection I should like to give expression to an idea which has occurred to me recently. I lived in solitude in the country and noticed how the monotony of a quiet life stimulates the creative mind. There are certain callings in our modern organization which entail such an isolated life without making a great claim on bodily and intellectual effort. I think of such occupations as the service in lighthouses and lightships. Would it not be possible to fill such places with young people who wish to think out scientific problems, especially of a mathematical or philosophical nature? Very few of such people have the opportunity during the most productive period of their lives to devote themselves undisturbed for any length of time to scientific problems. Even if a young person is lucky enough to obtain a scholarship for a short period he must endeavor to arrive as quickly as possible at definite conclusions. That cannot be of advantage in the pursuit of pure science. The young scientist who carries on an ordinary practical profession which maintains him is in a much better position—assuming of course that this profession leaves him with sufficient spare time and energy. In this way perhaps a greater number of creative individuals could be given an opportunity for mental development than is possible at present. In these times of economic depression and political upheaval such considerations seem to be worth attention.

Shall we worry over the fact that we are living in a time of danger and want? I think not. Man like every other animal is by nature indolent. If nothing spurs him on, then he will hardly think, and will behave from habit like an automaton.

I am no longer young and can, therefore, say, that as a child and as a young man I experienced that phase—when a young man thinks only about the trivialities of personal existence, and talks like his fellows and behaves like them. Only with difficulty can one see what is really behind such a conventional mask. For owing to habit and speech his real personality is, as it were, wrapped in cotton wool.

How different it is today! In the lightning flashes of our tempestuous times one sees human beings and things in their nakedness. Every nation and every human being reveal clearly their aims, powers and weaknesses, and also their passions. Routine becomes of no avail under the swift change of condition; conventions fall away like dry husks.

Men in their distress begin to think about the failure of economic practice and about the necessity of political combinations which are supernational. Only through perils and upheavals can Nations be brought to further developments. May the present upheavals lead to a better world.

Above and beyond this valuation of our time we have this further duty, the care for what is eternal and highest amongst our possessions, that which gives to life its import and which we wish to hand on to our children purer and richer than we received it from our forbears.

8

A Message to Intellectuals*

WE MEET TODAY, as intellectuals and scholars of many nationalities, with a deep and historic responsibility placed upon us. We have every reason to be grateful to our French and Polish colleagues whose initiative has assembled us here for a momentous objective: to use the influence of wise men in promoting peace and security throughout the world. This is the age-old problem with which Plato, as one of the first, struggled so hard: to apply reason and prudence to the solution of man's problems instead of yielding to atavist instincts and passions.

By painful experience we have learnt that rational thinking does not suffice to solve the problems of our social life. Penetrating research and keen scientific work have often had tragic implications for mankind, producing, on the one hand, inventions which liberated man from exhausting physical labor, making his life easier and richer; but on the other hand, introducing a grave restlessness into his life, making him a slave to his technological environment, and—most catastrophic of all—creating the means for his own mass destruction. This, indeed, is a tragedy of overwhelming poignancy!

However poignant that tragedy is, it is perhaps even more tragic that, while mankind has produced many scholars so

*   The following address was objected to by the Organizing Committee of the Intellectuals' Conference for Peace. The message was subsequently released to the press on August 29, 1948.

extremely successful in the field of science and technology, we have been for a long time so inefficient in finding adequate solutions to the many political conflicts and economic tensions which beset us. No doubt, the antagonism of economic interests within and among nations is largely responsible to a great extent for the dangerous and threatening condition in the world today. Man has not succeeded in developing political and economic forms of organization which would guarantee the peaceful coexistence of the nations of the world. He has not succeeded in building the kind of system which would eliminate the possibility of war and banish forever the murderous instruments of mass destruction.

We scientists, whose tragic destination has been to help in making the methods of annihilation more gruesome and more effective, must consider it our solemn and transcendent duty to do all in our power in preventing these weapons from being used for the brutal purpose for which they were invented. What task could possibly be more important for us? What social aim could be closer to our hearts? That is why this Congress has such a vital mission. We are here to take counsel with each other. We must build spiritual and scientific bridges linking the nations of the world. We must overcome the horrible obstacles of national frontiers.

In the smaller entities of community life, man has made some progress toward breaking down anti-social sovereignties. This is true, for example, of life within cities and, to a certain degree, even of society within individual states. In such communities tradition and education have had a moderating influence and have brought about tolerable relations among the peoples living within those confines. But in relations among separate states complete anarchy still prevails. I do not believe that we have made any genuine advance

in this area during the last few thousand years. All too frequently conflicts among nations are still being decided by
brutal power, by war. The unlimited desire for ever greater
power seeks to become active and aggressive wherever and
whenever the physical possibility offers itself.

Throughout the ages, this state of anarchy in international
affairs has inflicted indescribable suffering and destruction
upon mankind; again and again it has depraved the development of men, their souls and their well-being. For given time
it has almost annihilated whole areas.

However, the desire of nations to be constantly prepared
for warfare has, however, still other repercussions upon the
lives of men. The power of every state over its citizens has
grown steadily during the last few hundred years, no less in
countries where the power of the state has been exercised
wisely, than in those where it has been used for brutal tyranny. The function of the state to maintain peaceful and
ordered relations among and between its citizens has become
increasingly complicated and extensive largely because of the
concentration and centralization of the modern industrial
apparatus. In order to protect its citizens from attacks from
without a modern state requires a formidable, expanding
military establishment. In addition, the state considers it
necessary to educate its citizens for the possibilities of war, an
"education" not only corrupting to the soul and spirit of the
young, but also adversely affecting the mentality of adults. No
country can avoid this corruption. It pervades the citizenry
even in countries which do not harbor outspoken aggressive
tendencies. The state has thus become a modern idol whose
suggestive power few men are able to escape.

Education for war, however, is a delusion. The technological developments of the last few years have created a

completely new military situation. Horrible weapons have
been invented, capable of destroying in a few seconds huge
masses of human beings and tremendous areas of territory.
Since science has not yet found protection from these weapons, the modern state is no longer in a position to prepare
adequately for the safety of its citizens.

How, then, shall we be saved?

Mankind can only gain protection against the danger
of unimaginable destruction and wanton annihilation if a
supranational organization has alone the authority to produce or possess these weapons. It is unthinkable, however,
that nations under existing conditions would hand over
such authority to a supranational organization unless the
organization would have the legal right and duty to solve all
the conflicts which in the past have led to war. The functions
of individual states would be to concentrate more or less
upon internal affairs; in their relation with other states they
would deal only with issues and problems which are in no
way conducive to endangering international security.

Unfortunately, there are no indications that governments
yet realize that the situation in which mankind finds itself
makes the adoption of revolutionary measures a compelling
necessity. Our situation is not comparable to anything in the
past. It is impossible, therefore, to apply methods and measures which at an earlier age might have been sufficient. We
must revolutionize our thinking, revolutionize our actions,
and must have the courage to revolutionize relations among
the nations of the world. Clichés of yesterday will no longer
do today, and will, no doubt, be hopelessly out of date tomorrow. To bring this home to men all over the world is the
most important and most fateful social function intellectuals
have ever had to shoulder. Will they have enough courage

to overcome their own national ties to the extent that is necessary to induce the peoples of the world to change their deep-rooted national traditions in a most radical fashion?

A tremendous effort is indispensable. If it fails now, the supranational organization will be built later, but then it will have to be built upon the ruins of a large part of the now existing world. Let us hope that the abolition of the existing international anarchy will not need to be bought by a self-inflicted world catastrophe the dimensions of which none of us can possibly imagine. The time is terribly short. We must act now if we are to act at all.

9

Open Letter to the General Assembly of the United Nations

WE ARE CAUGHT in a situation in which every citizen of every country, his children, and his life's work, are threatened by the terrible insecurity which reigns in our world today. The progress of technological development has not increased the stability and the welfare of humanity. Because of our inability to solve the problem of international organization, it has actually contributed to the dangers which threaten peace and the very existence of mankind.

The delegates of fifty-five governments, meeting in the second General Assembly of the United Nations, undoubtedly will be aware of the fact that during the last two years—since the victory over the Axis powers—no appreciable progress has been made either toward the prevention of war or toward agreement in specific fields such as control of atomic energy and economic cooperation in the reconstruction of war-devastated areas.

The UN cannot be blamed for these failures. No international organization can be stronger than the constitutional powers given it, or than its component parts want it to be. As a matter of fact, the United Nations is an extremely important and useful institution provided the peoples and governments of the world realize that it is merely a transitional system toward the final goal, which is the establishment of a supranational authority vested with sufficient legislative and executive powers to keep the peace. The present impasse lies in the fact that there is no sufficient, reliable supranational authority. Thus the responsible leaders of all governments are obliged to act on the assumption of eventual war. Every step motivated by that assumption contributes to the general fear and distrust and hastens the final catastrophe. However strong national armaments may be, they do not create military security for any nation nor do they guarantee the maintenance of peace.

There can never be complete agreement on international control and the administration of atomic energy or on general disarmament until there is a modification of the traditional concept of national sovereignty. For as long as atomic energy and armaments are considered a vital part of national security no nation will give more than lip service to international treaties. Security is indivisible. It can be reached only when necessary guarantees of law and enforcement obtain everywhere, so that military security is no longer the problem of any single state. There is no compromise possible between preparation for war, on the one hand, and preparation of a world society based on law and order on the other.

Every citizen must make up his mind. If he accepts the premise of war, he must reconcile himself to the maintenance of troops in strategic areas like Austria and Korea; to the sending of troops to Greece and Bulgaria; to the accumulation of stockpiles of uranium by whatever means; to universal military training, to the progressive limitation of civil liberties. Above all, he must endure the consequences of military secrecy which is one of the worst scourges of our time and one of the greatest obstacles to cultural betterment.

If on the other hand every citizen realizes that the only

guarantee for security and peace in this atomic age is the constant development of a supranational government, then he will do everything in his power to strengthen the United Nations. It seems to me that every reasonable and responsible citizen in the world must know where his choice lies.

Yet the world at large finds itself in a vicious circle since the UN powers seem to be incapable of making up their minds on this score. The Eastern and Western blocs each attempt frantically to strengthen their respective power positions. Universal military training, Russian troops in Eastern Europe, United States control over the Pacific Islands, even the stiffening colonial policies of the Netherlands, Great Britain and France, atomic and military secrecy—are all part of the old familiar jockeying for position.

The time has come for the UN to strengthen its moral authority by bold decisions. First, the authority of the General Assembly must be increased so that the Security Council as well as all other bodies of the UN will be subordinated to it. As long as there is a conflict of authority between the Assembly and the Security Council, the effectiveness of the whole institution will remain necessarily impaired.

Second, the method of representation at the UN should be considerably modified. The present method of selection by government appointment does not leave any real freedom to the appointee. Furthermore, selection by governments cannot give the peoples of the world the feeling of being fairly and proportionately represented. The moral authority of the UN would be considerably enhanced if the delegates were elected directly by the people. Were they responsible to an electorate, they would have much more freedom to follow their consciences. Thus we could hope for more statesmen and fewer diplomats.

Third, the General Assembly should remain in session throughout the critical period of transition. By staying constantly on the job, the Assembly could fulfill two major tasks: first, it could take the initiative toward the establishment of a supranational order; second, it could take quick and effective steps in all those danger areas (such as currently exist on the Greek border) where peace is threatened.

The Assembly, in view of these high tasks, should not delegate its powers to the Security Council, especially while that body is paralyzed by the shortcomings of the veto provisions. As the only body competent to take the initiative boldly and resolutely, the UN must act with utmost speed to create the necessary conditions for international security by laying the foundations for a real world government.

Of course there will be opposition. It is by no means certain that the U.S.S.R.—which is often represented as the main antagonist to the idea of world government—would maintain its opposition if an equitable offer providing for real security were made. Even assuming that Russia is now opposed to the idea of world government, once she becomes convinced that world government is nonetheless in the making her whole attitude may change. She may then insist on only the necessary guarantees of equality before the law so as to avoid finding herself in perennial minority as in the present Security Council.

Nevertheless, we must assume that despite all efforts Russia and her allies may still find it advisable to stay out of such a world government. In that case—and only after all efforts have been made in utmost sincerity to obtain the cooperation of Russia and her allies—the other countries would have to proceed alone. It is of the utmost importance that this partial world government be very strong, comprising at least two-thirds of the major industrial and economic areas of the world. Such strength in itself would make it possible for the partial world government to abandon military secrecy and all the other practices born of insecurity.

Such a partial world government should make it clear from the beginning that its doors remain wide open to any nonmember—particularly Russia—for participation on the basis of complete equality. In my opinion, the partial world government should accept the presence of observers from nonmember governments at all its meetings and constitutional conventions.

In order to achieve the final aim—which is one world, and not two hostile worlds—such a partial world government must never act as an alliance against the rest of the world. The only real step toward world government is world government itself.

In a world government the ideological differences between the various component parts are of no grave consequence. I am convinced that the present difficulties between the U.S.A. and the U.S.S.R. are not due primarily to ideological differences. Of course, these ideological differences are a contributing element in an already serious tension. But I am convinced that even if the U.S.A. and Russia were both capitalist countries—or communist, or monarchist, for that matter—their rivalries, conflicting interests, and jealousies would result in strains similar to those existing between the two countries today.

The UN now and world government eventually must serve one single goal—the guarantee of the security, tranquillity, and the welfare of all mankind.

10

Dr. Einstein's Mistaken Notions

AN OPEN LETTER FROM SERGEI VAVILOV, A. N. FRUMKIN, A. F. JOFFE, AND N. N. SEMYONOV*

THE CELEBRATED PHYSICIST, Albert Einstein, is famed not only for his scientific discoveries; of late years he has paid much attention to social and political problems. He speaks over the radio and writes in the press. He is associated with a number of public organizations. Time and again he raised his voice in protest against the Nazi barbarians. He is an advocate of enduring peace, and has spoken against the threat of a new war, and against the ambition of the militarists to bring American science completely under their control.

Soviet scientists, and the Soviet people in general, are appreciative of the humanitarian spirit which prompts these activities of the scientist, although his position has not always been as consistent and clear-cut as might be desired. However,

* *Biographical Note:* Sergei Vavilov, a physicist specializing in the field of fluorescence, is President of the Academy of Sciences of the USSR. A. N. Frumkin, a colloid chemist of note, is Director of the Colloid-Electrochemical Institute of the Academy of Sciences in Moscow. A. F. Joffe is well known for his work on the behavior of crystals under water, and is Director of the Physico-Chemical Institute of the Academy in Leningrad. N. N. Semyonov, an authority on chemical kinetics, is Director of the Institute of Chemical Physics of the Academy in Moscow.

in some of Einstein's more recent utterances there have been aspects which seem to us not only mistaken, but positively prejudicial to the cause of peace which Einstein so warmly espouses.

We feel it our duty to draw attention to this, in order to clarify so important a question as to how most effectively to work for peace. It is from this point of view that the idea of a "world government" which Dr. Einstein has of late been sponsoring must be considered.

In the motley company of proponents of this idea, besides out-and-out imperialists who are using it as a screen for unlimited expansion, there are quite a number of intellectuals in the capitalist countries who are captivated by the plausibility of the idea, and who do not realize its actual implications. These pacifist and liberal-minded individuals believe that a "world government" would be an effective panacea against the world's evils and a guardian of enduring peace.

The advocates of a "world government" make wide use of the seemingly radical argument that in this atomic age state sovereignty is a relic of the past, that it is, as Spaak, the Belgian delegate, said in the UN General Assembly, an "old-fashioned" and even "reactionary" idea. It would be hard to imagine an allegation that is farther from the truth.

In the first place, the idea of a "world government" and "superstate" are by no means products of the atomic age. They are much older than that. They were mooted, for instance, at the time the League of Nations was formed.

Further, these ideas have never been progressive in these modern times. They are a reflection of the fact that the capitalist monopolies, which dominate the major industrial countries, find their own national boundaries too narrow.

They need a world-wide market, worldwide sources of raw materials, and worldwide spheres of capital investment. Thanks to their domination in political and administrative affairs, the monopoly interests of the big powers are in a position to utilize the machinery of government in their struggle for spheres of influence and their efforts economically and politically to subjugate other countries, to play the master in these countries as freely as in their own.

We know this very well from the past experience of our own country. Under tsarism, Russia, with her reactionary regime, which was servilely accommodating to the interests of capital, with her low-paid labor and vast natural resources, was an alluring morsel to foreign capitalists. French, British, Belgian and German firms battened on our country like birds of prey, earning profits which would have been inconceivable in their own countries. They chained tsarist Russia to the capitalist West with extortionate loans. Supported by funds obtained from foreign banks, the tsarist government brutally repressed the revolutionary movement, retarded the development of Russian science and culture, and instigated Jewish pogroms.

The Great October Socialist Revolution smashed the chains of economic and political dependence that bound our country to the world capitalist monopolies. The Soviet Government made our country for the first time a really free and independent state, promoted the progress of our Socialist economy, technology, science and culture at a speed hitherto unwitnessed in history, and turned our country into a reliable bulwark of international peace and security. Our people upheld their country's independence in the civil war, in the struggle against the intervention of a bloc of imperialist states, and in the great battles of the war against the Nazi invaders.

And now the proponents of a "world super-state" are asking us voluntarily to surrender this independence for the sake of a "world government," which is nothing but a flamboyant signboard for the world supremacy of the capitalist monopolies.

It is obviously preposterous to ask of us anything like that. And it is not only with regard to the Soviet Union that such a demand is absurd. After World War II, a number of countries succeeded in breaking away from the imperialist system of oppression and slavery. The peoples of these countries are working to consolidate their economic and political independence, debarring alien interference in their domestic affairs. Further, the rapid spread of the movement for national independence in the colonies and dependencies has awakened the national consciousness of hundreds of millions of people, who do not desire to remain in the status of slaves any longer.

The monopolies of the imperialist countries, having lost a number of profitable spheres of exploitation, and running the risk of losing more, are doing their utmost to deprive the nations that have escaped from their mastery of the state independence which they, the monopolies, find so irksome, and to prevent the genuine liberation of the colonies. With this purpose, the imperialists are resorting to the most diverse methods of military, political, economic and ideological warfare.

It is in accordance with this social behest that the ideologians of imperialism are endeavoring to discredit the very idea of national sovereignty. One of the methods they resort to is the advocacy of pretentious plans for a "world state," which will allegedly do away with imperialism, wars and national enmity, ensure the triumph of universal law, and so on.

The predatory appetites of the imperialist forces that are striving for world supremacy are thus disguised under the garb of a pseudo-progressive idea which appeals to certain intellectuals—scientists, writers and others—in the capitalist countries.

In an open letter which he addressed last September to the United Nations delegations, Dr. Einstein suggested a new scheme for limiting national sovereignty. He recommends that the General Assembly be reconstructed and converted into a permanently functioning world parliament endowed with greater authority than the Security Council, which, Einstein declares (repeating what the henchmen of American diplomacy are asserting day in and day out), is paralyzed by the veto right. The General Assembly, reconstructed in accordance with Dr. Einstein's plan, is to have final powers of decision, and the principle of the unanimity of the Great Powers is to be abandoned.

Einstein suggests that the delegates to the United Nations should be chosen by popular election and not appointed by their governments, as at present. At a first glance, this proposal may seem progressive and even radical. Actually, it will in no way improve the existing situation.

Let us picture to ourselves what elections to such a "world parliament" would mean in practice.

A large part of humanity still lives in colonial and dependent countries dominated by the governors, the troops, and the financial and industrial monopolies of a few imperialist powers. "Popular election" in such countries would in practice mean the appointment of delegates by the colonial administration or the military authorities. One does not have to go far for examples; one need only recall the parody of a referendum in Greece, which was carried out by her royalist-fascist rulers under the protection of British bayonets.

But things would be not much better in the countries where universal suffrage formally exists. In the bourgeois-democratic countries, where capital dominates, the latter resorts to thousands of tricks and devices to turn universal suffrage and freedom of ballot into a farce. Einstein surely knows that in the last Congressional elections in the United States only 39 per cent of the electorate went to the polls; he surely knows that millions of Negroes in the Southern states are virtually deprived of the franchise, or are forced, not infrequently under threat of lynching, to vote for their bitterest enemies, such as the late arch-reactionary and Negrophobe, Senator Bilbo.

Poll taxes, special tests and other devices are employed to rob millions of immigrants, migrant workers and poor farmers of the vote. We will not mention the widespread practice of purchasing votes, the role of the reactionary press, that powerful instrument for influencing the masses wielded by millionaire newspaper proprietors, and so forth.

All this shows what popular elections to a world parliament, as suggested by Einstein, would amount to under existing conditions in the capitalist world. Its composition would be no better than the present composition of the General Assembly. It would be a distorted reflection of the real sentiments of the masses, of their desire and hope for lasting peace.

As we know, in the General Assembly and the UN committees, the American delegation has a regular voting machine at its disposal, thanks to the fact that the overwhelming majority of the members of the UN are dependent on the United

States and are compelled to adapt their foreign policy to the requirements of Washington. A number of Latin-American countries, for instance, countries with single-crop agricultural systems, are bound hand and foot to the American monopolies, which determine the prices of their produce. Such being the case, it is not surprising that, under pressure of the American delegation, a mechanical majority has arisen in the General Assembly which votes in obedience to the orders of its virtual masters.

There are cases when American diplomacy finds it preferable to realize certain measures, not through the State Department, but under the flag of the United Nations. Witness the notorious Balkan committee or the commission appointed to observe the elections in Korea.

It is with the object of converting the UN into a branch of the State Department that the American delegation is forcing through the project for a "Little Assembly," which would in practice replace the Security Council, with its principle of unanimity of the Great Powers that is proving such an obstacle to the realization of imperialist schemes.

Einstein's suggestion would lead to the same result, and thus, far from promoting lasting peace and international cooperation, would only serve as a screen for an offensive against nations which have established regimes that prevent foreign capital from extorting its customary profits. It would further the unbridled expansion of American imperialism, and ideologically disarm the nations which insist upon maintaining their independence.

By the irony of fate, Einstein has virtually become a supporter of the schemes and ambitions of the bitterest foes of peace and international cooperation. He has gone so far in this direction as to declare in advance in his open letter that if the Soviet Union refuses to join his newfangled organization, other countries would have every right to go ahead without it, while leaving the door open for eventual Soviet participation in the organization as a member or as an "observer."

Essentially this proposal differs very little from the suggestions of frank advocates of American imperialism, however remote Dr. Einstein may be from them in reality. The sum and substance of these suggestions is that if UN cannot be converted into a weapon of United States policy, into a screen for imperialist schemes and designs, that organization should be wrecked and a new "international" organization formed in its place, without the Soviet Union and the new democracies.

Does Einstein not realize how fatal such plans would be to international security and international cooperation?

We believe that Dr. Einstein has entered a false and dangerous path; he is chasing the mirage of a "world state" in a world where different social, political and economic systems exist. Of course there is no reason why states with different social and economic structures should not cooperate economically and politically, provided that these differences are soberly faced. But Einstein is sponsoring a political fad which plays into the hands of the sworn enemies of sincere international cooperation and enduring peace. The course he is inviting the member states of the United Nations to adopt would lead not to greater international security, but to new international complications. It would benefit only the capitalist monopolies, for whom new international complications hold out the promise of more war contracts and more profits.

It is because we so highly esteem Einstein as an eminent scientist and as a man of public spirit who is striving to the best of his ability to promote the cause of peace, that we consider it our duty to speak with utter frankness and without diplomatic adornment.

A REPLY TO THE SOVIET SCIENTISTS

Four of my Russian colleagues have published a benevolent attack upon me in an open letter carried by the *New Times*. I appreciate the effort they have made and I appreciate even more the fact that they have expressed their point of view so candidly and straightforwardly. To act intelligently in human affairs is only possible if an attempt is made to understand the thoughts, motives, and apprehensions of one's opponent so fully that one can see the world through his eyes. All well-meaning people should try to contribute as much as possible to improving such mutual understanding. It is in this spirit that I should like to ask my Russian colleagues and any other reader to accept the following answer to their letter. It is the reply of a man who anxiously tries to find a feasible solution without having the illusion that he himself knows "the truth" or "the right path" to follow.

If in the following I shall express my views somewhat dogmatically, I do it only for the sake of clarity and simplicity.

Although your letter, in the main, is clothed in an attack upon the non-socialistic foreign countries, particularly the United States, I believe that behind the aggressive front there lies a defensive mental attitude which is nothing else but the trend towards an almost unlimited isolationism. The escape into isolationism is not difficult to understand if one realizes what Russia has suffered at the hands of foreign countries during the last three decades—the German invasions with planned mass murder of the civilian population, foreign interventions during the civil war, the systematic campaign of calumnies in the western press, the support of Hitler as an alleged tool to fight Russia. However understandable this desire for isolation may be, it remains no less disastrous to Russia and to all other nations; I shall say more about it later on.

The chief object of your attack against me concerns my support of "world government." I should like to discuss this important problem only after having said a few words about the antagonism between socialism and capitalism; for your attitude on the significance of this antagonism seems to dominate completely your views on international problems. If the socio-economic problem is considered objectively, it appears as follows: technological development has led to increasing centralization of the economic mechanism. It is this development which is also responsible for the fact that economic power in all widely industrialized countries has become concentrated in the hands of relatively few. These people, in capitalist countries, do not need to account for their actions to the public as a whole; they must do so in socialist countries in which they are civil servants similar to those who exercise political power.

I share your view that a socialist economy possesses advantages which definitely counterbalance its disadvantages whenever the management lives up, at least to some extent, to adequate standards. No doubt, the day will come when all nations (as far as such nations still exist) will be grateful to Russia for having demonstrated, for the first time, by vigorous action the practical possibility of planned economy in spite of exceedingly great difficulties. I also believe that capitalism, or, we should say, the system of free enterprise, will prove unable to check unemployment, which will become increasingly chronic because of technological progress, and unable to maintain a healthy balance between production and the purchasing power of the people.

On the other hand we should not make the mistake of blaming capitalism for all existing social and political evils, and of assuming that the very establishment of socialism would be able to cure all the social and political ills of humanity. The danger of such a belief lies, first, in the fact that it encourages fanatical intolerance on the part of all the "faithful" by making a possible social method into a type of church which brands all those who do not belong to it as traitors or as nasty evildoers. Once this stage has been reached, the ability to understand the convictions and actions of the "unfaithful" vanishes completely. You know, I am sure, from history how much unnecessary suffering such rigid beliefs have inflicted upon mankind.

Any government is in itself an evil insofar as it carries within it the tendency to deteriorate into tyranny. However, except for a very small number of anarchists, everyone of us is convinced that civilized society cannot exist without a government. In a healthy nation there is a kind of dynamic balance between the will of the people and the government, which prevents its degeneration into tyranny. It is obvious that the danger of such deterioration is more acute in a country in which the government has authority not only over the armed forces but also over all the channels of education and information as well as over the economic existence of every single citizen. I say this merely to indicate that socialism as such cannot be considered the solution to all social problems but merely as a framework within which such a solution is possible.

What has surprised me most in your general attitude,

expressed in your letter, is the following aspect: You are such passionate opponents of anarchy in the economic sphere, and yet equally passionate advocates of anarchy, e.g., unlimited sovereignty, in the sphere of international politics. The proposition to curtail the sovereignty of individual states appears to you in itself reprehensible, as a kind of violation of a natural right. In addition, you try to prove that behind the idea of curtailing sovereignty the United States is hiding her intention of economic domination and exploitation of the rest of the world without going to war. You attempt to justify this indictment by analyzing in your fashion the individual actions of this government since the end of the last war. You attempt to show that the Assembly of the United Nations is a mere puppet show controlled by the United States and hence the American capitalists.

Such arguments impress me as a kind of mythology; they are not convincing. They make obvious, however, the deep estrangement among the intellectuals of our two countries which is the result of a regrettable and artificial mutual isolation. If a free personal exchange of views should be made possible and should be encouraged, the intellectuals, possibly more than anyone else, could help to create an atmosphere of mutual understanding between the two nations and their problems. Such an atmosphere is a necessary prerequisite for the fruitful development of political cooperation. However, since for the time being we depend upon the cumbersome method of "open letters" I shall want to indicate briefly my reaction to your arguments.

Nobody would want to deny that the influence of the economic oligarchy upon all branches of our public life is very powerful. This influence, however, should not be overestimated. Franklin Delano Roosevelt was elected president in spite of desperate opposition by these very powerful groups and was reelected three times; and this took place at a time when decisions of great consequence had to be made.

Concerning the policies of the American Government since the end of the war, I am neither willing, nor able, nor entitled to justify or explain them. It cannot be denied, however, that the suggestions of the American Government with regard to atomic weapons represented at least an attempt towards the creation of a supranational security organization. If they were not acceptable, they could at least have served as a basis of discussion for a real solution of the problems of international security. It is, indeed, the attitude of the Soviet Government, that was partly negative and partly dilatory, which has made it so difficult for well-meaning people in this country to use their political influence as they would have wanted, and to oppose the "war mongers." With regard to the influence of the United States upon the United Nations Assembly, I wish to say that, in my opinion, it stems not only from the economic and military power of the United States but also from the efforts of the United States and the United Nations to lead toward a genuine solution of the security problem.

Concerning the controversial veto power, I believe that the efforts to eliminate it or to make it ineffective have their primary cause less in specific intentions of the United States than in the manner in which the veto privilege has been abused.

Let me come now to your suggestion that the policy of the United States seeks to obtain economic domination and exploitation of other nations. It is a precarious undertaking to say anything reliable about aims and intentions. Let us rather examine the objective factors involved. The United

States is fortunate in producing all the important industrial products and foods in her own country, in sufficient quantities. The country also possesses almost all important raw materials. Because of her tenacious belief in "free enterprise" she cannot succeed in keeping the purchasing power of the people in balance with the productive capacity of the country. For these very same reasons there is a constant danger that unemployment will reach threatening dimensions.

Because of these circumstances the United States is compelled to emphasize her export trade. Without it, she could not permanently keep her total productive machinery fully utilized. These conditions would not be harmful if the exports were balanced by imports of about the same value. Exploitation of foreign nations would then consist in the fact that the labor value of imports would considerably exceed that of exports. However, every effort is being made to avoid this, since almost every import would make a part of the productive machinery idle.

This is why foreign countries are not able to pay for the export commodities of the United States, payment which, in the long run, would indeed be possible only through imports by the latter. This explains why a large portion of all the gold has come to the United States. On the whole, this gold cannot be utilized except for the purchase of foreign commodities, which because of the reasons already stated, is not practicable. There it lies, this gold, carefully protected against theft, a monument to governmental wisdom and to economic science! The reasons which I have just indicated make it difficult for me to take the alleged exploitation of the world by the United States very seriously.

However, the situation just described has a serious political facet. The United States, for the reasons indicated, is compelled to ship part of its production to foreign countries. These exports are financed through loans which the United States is granting foreign countries. It is, indeed, difficult to imagine how these loans will ever be repaid. For all practical purposes, therefore, these loans must be considered gifts which may be used as weapons in the arena of power politics. In view of the existing conditions and in view of the general characteristics of human beings, this, I frankly admit, represents a real danger. Is it not true, however, that we have stumbled into a state of international affairs which tends to make every invention of our minds and every material good into a weapon and, consequently, into a danger for mankind?

This question brings us to the most important matter, in comparison to which everything else appears insignificant indeed. We all know that power politics, sooner or later, necessarily leads to war, and that war, under present circumstances, would mean a mass destruction of human beings and material goods, the dimensions of which are much, much greater than anything that has ever before happened in history.

Is it really unavoidable that, because of our passions and our inherited customs, we should be condemned to annihilate each other so thoroughly that nothing would be left over which would deserve to be conserved? Is it not true that all the controversies and differences of opinion which we have touched upon in our strange exchange of letters are insignificant pettinesses compared to the danger in which we all find ourselves? Should we not do everything in our power to eliminate the danger which threatens all nations alike?

If we hold fast to the concept and practice of unlimited sovereignty of nations it only means that each country reserves the right for itself of pursuing its objectives through warlike means. Under the circumstances, every nation must be prepared for that possibility; this means it must try with all its might to be superior to anyone else. This objective will dominate more and more our entire public life and will poison our youth long before the catastrophe is itself actually upon us. We must not tolerate this, however, as long as we still retain a tiny bit of calm reasoning and human feelings.

This alone is on my mind in supporting the idea of "World Government," without any regard to what other people may have in mind when working for the same objective. I advocate world government because I am convinced that there is no other possible way of eliminating the most terrible danger in which man has ever found himself. The objective of avoiding total destruction must have priority over any other objective.

I am sure you are convinced that this letter is written with all the seriousness and honesty at my command; I trust you will accept it in the same spirit.

11

For an Organization of Intellectual Workers

I CONSIDER IT IMPORTANT, indeed urgently necessary, for intellectual workers to get together, both to protect their own economic status and also, generally speaking, to secure their influence in the political field.

On the first-mentioned, the economic side, the working class may serve us as a model: they have succeeded, at least to some extent, in protecting their economic interests. We can learn from them too how this problem can be solved by the method of organization. And also, we can learn from them what is our gravest danger, which we ourselves must seek to avoid: the weakening through inner dissensions, which, when things reach that point, make cooperation difficult and result in quarrels between the constituent groups.

But again, we can also learn from the workers that limitation to immediate economic aims, to the exclusion of all political goals and effective action will not suffice either. In this respect, the working classes in this country have only begun their development. It is inevitable, considering the progressive centralization of production, that the economic and the political struggle should become more and more closely interwoven, the political factor continually growing in significance in the process. In the meantime the intellectual worker, due to his lack of organization, is less well protected against arbitrariness and exploitation than a member of any other calling.

But intellectual workers should unite, not only in their own interest but also and no less importantly in the interest of society as a whole. For division among intellectuals has been partly to blame for the fact that the special parts and the experience which are the birthright of these groups have so seldom been made available for political aims. In their room political ambition and desire for profit almost exclusively determine events, instead of professional knowledge and judgment based upon objective thinking.

An organization of intellectual workers can have the greatest significance for society as a whole by influencing public opinion through publicity and education. Indeed it is its proper task to defend academic freedom, without which a healthy development of democracy is impossible.

An outstandingly important task for an organization of intellectual workers at the present moment is to fight for the establishment of a supranational political force as a protection against fresh wars of aggression. It seems to me that the working out with a view to selection of a particular plan for an international government should not, at the present moment, be our chief aim. For if there existed, among the majority of citizens, the firm intention of establishing international security, the technique of giving shape to such an instrument would not present an all-too-difficult problem. What is lacking in the majority is the conviction, founded on clear thinking, that there is no other means of permanently avoiding catastrophes like the present one. In the organization and promotion of enlightenment on this subject, I see the most important service which an organization of intellectual workers can perform at this historic moment. Only by means of setting energetically about such a task can an organization like the one here planned achieve inward strength and outward influence.

12

"Was Europe a Success?"

THE HUMANITARIAN IDEAL of Europe appears indeed to be unalterably bound up with the free expression of opinion, to some extent with the free-will of the individual, with the effort toward objectivity in thought without consideration of mere utility, and with the encouragement of differences in the realm of mind and taste. These requirements and ideals comprise the nature of the European spirit. One cannot establish with reason the worth of these values and maxims, for they are matters of fundamental principle in the approach to life and are points of departure which can only be affirmed or denied by emotion. I only know that I affirm them with my whole soul, and would find it intolerable to belong to a society which consistently denied them.

I do not share the pessimism of those who believe that full intellectual growth is dependent on the foundation of open or concealed slavery. That may be true for eras of primitive technical development, where the production of the necessaries of life requires physical work by a majority of the people to the point of total exhaustion. In our time of high technical development, with a reasonably equitable division of labor and adequate provisions for all, the individual would have both time and strength to participate receptively and productively in the finest intellectual and artistic efforts his abilities and inclinations allowed. Unfortunately nothing approaching such conditions exist in our society. But everyone devoted to the specific European ideals will do his utmost to achieve aims of whose desirability and practicability an increasing number of right-minded persons are convinced.

Is it justifiable to set aside for a time the principles of individual freedom in deference to the high endeavor to improve economic organization? A fine and shrewd Russian scholar very skilfully defended this point of view to me in comparing the success of compulsion and terror—at least at the outset—in a functioning Russian Communism with the failure of German Social Democracy after the war. He did not convince me. No purpose is so high that unworthy methods in achieving it can be justified in my eyes. Violence sometimes may have cleared away obstructions quickly, but it never has proved itself creative.

13

At a Gathering for Freedom of Opinion

WE HAVE COME here today to defend the freedom of opinion guaranteed by the Constitution of the United States, and
also in defense of the freedom of teaching. By the same token
we wish to draw the attention of intellectual workers to the
great danger that now menaces these liberties.

How is such a thing possible? Why is the danger more
menacing than in years gone by? The centralization of production has brought about a concentration of productive
capital in the hands of a relatively small number of the citizens
of the land. This small group exerts an overwhelming domination over the institutions for the education of our youth as
well as over the great newspapers of the country. At the same
time it wields enormous influences on the government. This
in itself is already sufficient to constitute a serious menace to
the intellectual freedom of the nation. But there is the additional fact that this process of economic concentration has
given birth to a problem previously unknown—permanent
unemployment for part of those who are able to work. The
federal government is endeavoring to solve this problem by
systematic control over economic processes—that is to say, by
a limitation of the so-called free interplay of the fundamental
economic forces of supply and demand.

But circumstances are stronger than men. The dominant
economic minority, heretofore autonomous and responsible
to no one, has placed itself in opposition to this limitation of

its freedom of action, demanded for the good of the whole people. For its defense this minority is resorting to every known legal method at its disposal. We need not, therefore, be surprised that they are using their preponderant influence on the schools and the press to prevent youth from being enlightened on this problem which is so vital to the sound and peaceful development of life in this country.

It is for this reason that of late we have had to witness repeatedly the dismissal of worthy university teachers against the will of their colleagues, actions of which the press has informed the public but inadequately. It is also to the pressure of this economically dominant minority that we owe the unhappy institution of the teacher's oath, which is meant to diminish the freedom of teaching. I need not dwell on the point that freedom of teaching and of opinion in book or press is the foundation for the sound and natural development of any people. The lessons of history—especially the very latest chapters—are all too plain on this score. It is the bounden duty of everyone to stand with every ounce of energy for the preservation and enhancement of these liberties and to exert all possible influence in keeping public opinion aware of the existing danger.

These difficulties can be solved only when our great economic problem is solved in a democratic manner; but the groundwork for such a solution must be prepared by preserving freedom of expression. This, moreover, is also the only method by which the worst damage can be prevented.

Let all of us therefore summon our strength. Let us be tirelessly on guard, lest it be said later of the intellectual elite of this land: Timidly and without a struggle they surrendered the heritage handed down to them by their forefathers—a heritage of which they were not worthy.

14

Atomic War or Peace

I

THE RELEASE OF atomic energy has not created a new problem. It has merely made more urgent the necessity of solving an existing one. One could say that it has affected us quantitatively, not qualitatively. As long as there are sovereign nations possessing great power, war is inevitable. That is not an attempt to say when it will come, but only that it is sure to come. That was true before the atomic bomb was made. What has been changed is the destructiveness of war.

I do not believe that civilization will be wiped out in a war fought with the atomic bomb. Perhaps two-thirds of the people of the earth might be killed. But enough men capable of thinking, and enough books, would be left to start again, and civilization could be restored.

I do not believe that the secret of the bomb should be given to the United Nations Organization. I do not believe it should be given to the Soviet Union. Either course would be like a man with capital, and wishing another man to work with him on some enterprise, starting out by simply giving that man half of his money. The other man might choose to start a rival enterprise, when what is wanted is his cooperation. The secret of the bomb should be committed to a world government, and the United States should immediately announce its readiness to give it to a world government. This government should be founded by the United States, the Soviet Union and Great Britain, the only three powers with great military strength. All three of them should commit to this world government all of their military strength. The fact that there are only three nations with great military power should make it easier, rather than harder, to establish such a government.

Since the United States and Great Britain have the secret of the atomic bomb and the Soviet Union does not, they should invite the Soviet Union to prepare and present the first draft of a constitution of the proposed world government. That will help dispel the distrust of the Russians, which they already feel because the bomb is being kept a secret chiefly to prevent their having it. Obviously the first draft would not be the final one, but the Russians should be made to feel that the world government will assure them their security.

It would be wise if this constitution were to be negotiated by a single American, a single Briton and a single Russian. They would have to have advisers, but these advisers should only advise when asked. I believe three men can succeed in writing a workable constitution acceptable to them all. Six or seven men, or more, probably would fail. After the three great powers have drafted a constitution and adopted it, the smaller nations should be invited to join the world government. They should be free to stay out, and though they should feel perfectly secure in staying out, I am sure they would wish to join. Naturally they should be entitled to propose changes in the constitution as drafted by the Big Three. But the Big Three should go ahead and organize the world government, whether the smaller nations join or not.

The power of this world government would be over all military matters, and there need be only one further power. That is to interfere in countries where a minority is oppressing a majority, and so is creating the kind of instability that leads to war. Conditions such as exist in Argentina and Spain should be dealt with. There must be an end to the concept of non-intervention, for to end it is part of keeping the peace.

The establishment of this world government must not have to wait until the same conditions of freedom are to be found in all three of the great powers. While it is true that in the Soviet Union the minority rules, I do not consider that internal conditions there are of themselves a threat to world peace. One must bear in mind that the people in Russia did not have a long political education, and changes to improve Russian conditions had to be carried through by a minority for the reason that there was no majority capable of doing it. If I had been born a Russian, I believe I could have adjusted myself to this situation.

It should not be necessary, in establishing a world government with a monopoly of military authority, to change the structure of the three great powers. It would be for the three individuals who draft the constitution to devise ways for their different structures to be fitted together for collaboration.

Do I fear the tyranny of a world government? Of course I do. But I fear still more the coming of another war or wars. Any government is certain to be evil to some extent. But a world government is preferable to the far greater evil of wars, particularly with their intensified destructiveness. If such a world government is not established by a process of agreement, I believe it will come anyway, and in a much more dangerous form. For war or wars will end in one power being supreme and dominating the rest of the world by its overwhelming military strength.

Now we have the atomic secret, we must not lose it, and that is what we should risk doing, if we give it to the United

Nations Organization or to the Soviet Union. But we must make it clear as quickly as possible that we are not keeping the bomb a secret for the sake of our power, but in the hope of establishing peace through a world government, and we will do our utmost to bring this world government into being.

I appreciate that there are persons who favor a gradual approach to world government, even though they approve of it as the ultimate objective. The trouble with taking little steps, one at a time, in the hope of reaching the ultimate goal, is that while they are being taken, we continue to keep the bomb without making our reason convincing to those who do not have it. That of itself creates fear and suspicion, with the consequence that the relations of rival sovereignties deteriorate dangerously. So while persons who take only a step at a time may think they are approaching world peace, they actually are contributing by their slow pace to the coming of war. We have no time to spend in this way. If war is to be averted, it must be done quickly.

We shall not have the secret very long. I know it is argued that no other country has money enough to spend on the development of the atomic bomb, which assures us the secret for a long time. It is a mistake often made in this country to measure things by the amount of money they cost. But other countries which have the materials and the men and care to apply them to the work of developing atomic power can do so, for men and materials and the decision to use them, and not money, are all that are needed.

I do not consider myself the father of the release of atomic energy. My part in it was quite indirect. I did not, in fact, foresee that it would be released in my time. I believed only that it was theoretically possible. It became practical through the accidental discovery of chain reaction, and this was not something I could have predicted. It was discovered by Hahn in Berlin, and he himself misinterpreted what he discovered. It was Lize Meitner who provided the correct interpretation, and escaped from Germany to place the information in the hands of Niels Bohr.

I do not believe that a great era of atomic science is to be assured by organizing science, in the way large corporations are organized. One can organize to apply a discovery already made, but not to make one. Only a free individual can make a discovery. There can be a kind of organizing by which scientists are assured their freedom and proper conditions of work. Professors of science in American universities, for instance, should be relieved of some of their teaching so as to have time for more research. Can you imagine an organization of scientists making the discoveries of Charles Darwin?

Nor do I believe that the vast private corporations of the United States are suitable to the needs of these times. If a visitor should come to this country from another planet, would he not find it strange that in this country so much power is permitted to private corporations without their having commensurate responsibility? I say this to stress that the American government must keep the control of atomic energy, not because socialism is necessarily desirable, but because atomic energy was developed by the government, and it would be unthinkable to turn over this property of the people to any individuals or groups of individuals. As to socialism, unless it is international to the extent of producing world government which controls all military power, it might more easily lead to wars than capitalism, because it represents a still greater concentration of power.

To give any estimate when atomic energy can be applied to constructive purposes is impossible. What now is known is only how to use a fairly large quantity of uranium. The use of small quantities, sufficient, say, to operate a car or an airplane, so far is impossible, and one cannot predict when it will be achieved. No doubt, it will be achieved, but nobody can say when. Nor can one predict when materials more common than uranium can be used to supply atomic energy. Presumably all materials used for this purpose will be among the heavier elements of high atomic weight. Those elements are relatively scarce due to their lesser stability. Most of these materials may have already disappeared by radioactive disintegration. So though the release of atomic energy can be, and no doubt will be, a great boon to mankind, that may not be for some time.

I myself do not have the gift of explanation with which I am able to persuade large numbers of people of the urgency of the problems the human race now faces. Hence I should like to commend someone who has this gift of explanation, Emery Reves, whose book, *The Anatomy of the Peace*, is intelligent, clear, brief, and, if I may use the abused term, dynamic on the topic of war and need for world government.

Since I do not foresee that atomic energy is to be a great boon for a long time, I have to say that for the present it is a menace. Perhaps it is well that it should be. It may intimidate the human race to bring order into its international affairs, which, without the pressure of fear, it undoubtedly would not do.

II

Since the completion of the first atomic bomb nothing has been accomplished to make the world more safe from war, while much has been done to increase the destructiveness of war. I am not able to speak from any firsthand knowledge about the development of the atomic bomb, since I do not work in this field. But enough has been said by those who do to indicate that the bomb has been made more effective. Certainly the possibility can be envisaged of building a bomb of far greater size, capable of producing destruction over a larger area. It also is credible that an extensive use could be made of radioactivated gases which would spread over a wide region, causing heavy loss of life without damage to buildings.

I do not believe it is necessary to go on beyond these possibilities to contemplate a vast extension of bacteriological warfare. I am skeptical that this form presents dangers comparable with those of atomic warfare. Nor do I take into account a danger of starting a chain reaction of a scope great enough to destroy part or all of this planet. I dismiss this on the ground that if it could happen from a man-made atomic explosion it would already have happened from the action of the cosmic rays which are continually reaching the earth's surface.

But it is not necessary to imagine the earth being destroyed like a nova by a stellar explosion to understand vividly the growing scope of atomic war and to recognize that unless another war is prevented it is likely to bring destruction on a scale never before held possible and even now hardly conceived, and that little civilization would survive it.

In the first two years of the atomic era another phenomenon is to be noted. The public, having been warned of the horrible nature of atomic warfare, has done nothing about it, and to a large extent has dismissed the warning from its consciousness. A danger that cannot be averted had perhaps better be forgotten; or a danger against which every possible precaution has been taken also had probably better be forgotten. That is, if the United States had dispersed its industries and decentralized its cities, it might be reasonable for people to forget the peril they face.

I should say parenthetically that it is well that this country has not taken these precautions, for to have done so would make atomic war still more probable, since it would convince the rest of the world that we are resigned to it and are preparing for it. But nothing has been done to avert war, while much has been done to make atomic war more horrible; so there is no excuse for ignoring the danger.

I say that nothing has been done to avert war since the completion of the atomic bomb, despite the proposal for supranational control of atomic energy put forward by the United States in the United Nations. This country has made only a conditional proposal, and on conditions which the Soviet Union is now determined not to accept. This makes it possible to blame the failure on the Russians.

But in blaming the Russians the Americans should not ignore the fact that they themselves have not voluntarily renounced the use of the bomb as an ordinary weapon in the time before the achievement of supranational control, or if supranational control is not achieved. Thus they have fed the fear of other countries that they consider the bomb a legitimate part of their arsenal so long as other countries decline to accept their terms for supranational control.

Americans may be convinced of their determination not to launch an aggressive or preventive war. So they may believe it is superfluous to announce publicly that they will not a second time be the first to use the atomic bomb. But this country has been solemnly invited to renounce the use of the bomb—that is, to outlaw it—and has declined to do so unless its terms for supranational control are accepted.

I believe this policy is a mistake. I see a certain military gain from not renouncing the use of the bomb in that this may be deemed to restrain another country from starting a war in which the United States might use it. But what is gained in one way is lost in another. For an understanding over the supranational control of atomic energy has been made more remote. That may be no military drawback so long as the United States has the exclusive use of the bomb. But the moment another country is able to make it in substantial quantities, the United States loses greatly through the absence of an international agreement, because of the vulnerability of its concentrated industries and its highly developed urban life.

In refusing to outlaw the bomb while having the monopoly of it, this country suffers in another respect, in that it fails to return publicly to the ethical standards of warfare formally accepted previous to the last war. It should not be forgotten that the atomic bomb was made in this country as a preventive measure; it was to head off its use by the Germans, if they discovered it. The bombing of civilian centers was initiated by the Germans and adopted by the Japanese. To it the Allies responded in kind—as it turned out, with greater effectiveness—and they were morally justified in doing so. But now, without any provocation, and without the justification of reprisal or retaliation, a refusal to outlaw the use of the bomb save in reprisal is making a political purpose of its possession. This is hardly pardonable.

I am not saying that the United States should not manufacture and stockpile the bomb, for I believe that it must do so; it must be able to deter another nation from making an atomic attack when it also has the bomb. But deterrence should be the only purpose of the stockpile of bombs. In the same way I believe that the United Nations should have the atomic bomb when it is supplied with its own armed forces and weapons. But it too should have the bomb for the sole purpose of deterring an aggressor or rebellious nations from making an atomic attack. It should not use the atomic bomb on its own initiative any more than the United States or any other power should do so. To keep a stockpile of atomic bombs without promising not to initiate its use is exploiting the possession of the bombs for political ends. It may be that the United States hopes in this way to frighten the Soviet Union into accepting supranational control of atomic energy. But the creation of fear only heightens antagonism and increases the danger of war. I am of the opinion that this policy has detracted from the very real virtue in the offer of supranational control of atomic energy.

We have emerged from a war in which we had to accept the degradingly low ethical standards of the enemy. But instead of feeling liberated from his standards, and set free to restore the sanctity of human life and the safety of noncombatants, we are in effect making the low standards of the enemy in the last war our own for the present. Thus we are starting toward another war degraded by our own choice.

It may be that the public is not fully aware that in another war atomic bombs will be available in large quantities. It may measure the dangers in the terms of the three bombs exploded before the end of the last war. The public also may not appreciate that, in relation to the damage inflicted, atomic bombs already have become the most economical form of destruction that can be used on the offensive. In another war the bombs will be plentiful and they will be comparatively cheap. Unless there is a determination not to use them that is far stronger than can be noted today among American political and military leaders, and on the part of the public itself, atomic warfare will be hard to avoid. Unless Americans come to recognize that they are not stronger in the world because they have the bomb, but weaker because of their vulnerability to atomic attack, they are not likely to conduct their policy at Lake Success or in their relations with Russia in a spirit that furthers the arrival at an understanding.

But I do not suggest that the American failure to outlaw the use of the bomb except in retaliation is the only cause of the absence of an agreement with the Soviet Union over atomic control. The Russians have made it clear that they will do everything in their power to prevent a supranational regime from coming into existence. They not only reject it in the range of atomic energy: they reject it sharply on principle, and thus have spurned in advance any overture to join a limited world government.

Mr. Gromyko has rightly said that the essence of the American atomic proposal is that national sovereignty is not compatible with the atomic era. He declares that the Soviet Union cannot accept this thesis. The reasons he gives are obscure, for they quite obviously are pretexts. But what seems to be true is that the Soviet leaders believe they cannot preserve the social structure of the Soviet state in a supranational regime. The Soviet government is determined to maintain its present social structure, and the leaders of Russia, who hold their great power through the nature of that structure, will spare no effort to prevent a supranational regime from coming into existence, to control atomic energy or anything else.

The Russians may be partly right about the difficulty of

retaining their present social structure in a supranational regime, though in time they may be brought to see that this is a far lesser loss than remaining isolated from a world of law. But at present they appear to be guided by their fears, and one must admit that the United States has made ample contributions to these fears, not only as to atomic energy but in many other respects. Indeed this country has conducted its Russian policy as though it were convinced that fear is the greatest of all diplomatic instruments.

That the Russians are striving to prevent the formation of a supranational security system is no reason why the rest of the world should not work to create one. It has been pointed out that the Russians have a way of resisting with all their arts what they do not wish to have happen; but once it happens, they can be flexible and accommodate themselves to it. So it would be well for the United States and other powers not to permit the Russians to veto an attempt to create supranational security. They can proceed with some hope that once the Russians see they cannot prevent such a regime they may join it.

So far the United States has shown no interest in preserving the security of the Soviet Union. It has been interested in its own security, which is characteristic of the competition which marks the conflict for power between sovereign states. But one cannot know in advance what would be the effect on Russian fears if the American people forced their leaders to pursue a policy of substituting law for the present anarchy of international relations. In a world of law, Russian security would be equal to our own, and for the American people to espouse this wholeheartedly, something that should be possible under the workings of democracy, might work a kind of miracle in Russian thinking.

At present the Russians have no evidence to convince them that the American people are not contentedly supporting a policy of military preparedness which they regard as a policy of deliberate intimidation. If they had evidences of a passionate desire by Americans to preserve peace in the one way it can be maintained, by a supranational regime of law, this would upset Russian calculations about the peril to Russian security in current trends of American thought. Not until a genuine, convincing offer is made to the Soviet Union, backed by an aroused American public, will one be entitled to say what the Russian response would be.

It may be that the first response would be to reject the world of law. But if from that moment it began to be clear to the Russians that such a world was coming into existence without them, and that their own security was being increased, their ideas necessarily would change.

I am in favor of inviting the Russians to join a world government authorized to provide security, and if they are unwilling to join, to proceed to establish supranational security without them. Let me admit quickly that I see great peril in such a course. If it is adopted it must be done in a way to make it utterly clear that the new regime is not a combination of power against Russia. It must be a combination that by its composite nature will greatly reduce the chances of war. It will be more diverse in its interests than any single state, thus less likely to resort to aggressive or preventive war. It will be larger, hence stronger than any single nation. It will be geographically much more extensive, and thus more difficult to defeat by military means. It will be dedicated to supranational security, and thus escape the emphasis on national supremacy which is so strong a factor in war.

If a supranational regime is set up without Russia, its

service to peace will depend on the skill and sincerity with which it is done. Emphasis should always be apparent on the desire to have Russia take part. It must be clear to Russia, and no less so to the nations comprising the organization, that no penalty is incurred or implied because a nation declines to join. If the Russians do not join at the outset, they must be sure of a welcome when they do decide to join. Those who create the organization must understand that they are building with the final objective of obtaining Russian adherence.

These are abstractions, and it is not easy to outline the specific lines a partial world government must follow to induce the Russians to join. But two conditions are clear to me: the new organization must have no military secrets; and the Russians must be free to have observers at every session of the organization, where its new laws are drafted, discussed, and adopted, and where its policies are decided. That would destroy the great factory of secrecy where so many of the world's suspicions are manufactured.

It may affront the military-minded person to suggest a regime that does not maintain any military secrets. He has been taught to believe that secrets thus divulged would enable a war-minded nation to seek to conquer the earth. (As to the so-called secret of the atomic bomb, I am assuming the Russians will have this through their own efforts within a short time.) I grant there is a risk in not maintaining military secrets. If a sufficient number of nations have pooled their strength they can take this risk, for their security will be greatly increased. And it can be done with greater assurance because of the decrease of fear, suspicion, and distrust that will result. The tensions of the increasing likelihood of war in a world based on sovereignty would be replaced by the relaxation of the growing confidence in peace. In time this might so allure the Russian people that their leaders would mellow in their attitude toward the West.

Membership in a supranational security system should not, in my opinion, be based on any arbitrary democratic standards. The one requirement from all should be that the representatives to supranational organization—assembly and council—must be elected by the people in each member country through a secret ballot. These representatives must represent the people rather than any government—which would enhance the pacific nature of the organization.

To require that other democratic criteria be met is, I believe, inadvisable. Democratic institutions and standards are the result of historic developments to an extent not always appreciated in the lands which enjoy them. Setting arbitrary standards sharpens the ideological differences between the Western and Soviet systems.

But it is not the ideological differences which now are pushing the world in the direction of war. Indeed, if all the Western nations were to adopt socialism, while maintaining their national sovereignty, it is quite likely that the conflict for power between East and West would continue. The passion expressed over the economic systems of the present seems to me quite irrational. Whether the economic life of America should be dominated by relatively few individuals, as it is, or these individuals should be controlled by the state, may be important, but it is not important enough to justify all the feelings that are stirred up over it.

I should wish to see all the nations forming the supranational state pool all their military forces, keeping for themselves only local police. Then I should like to see these forces commingled and distributed as were the regiments of the former Austro-Hungarian Empire. There it was appreciated that the men and officers of one region would serve the purposes of empire better by not being stationed exclusively in their own provinces, subject to local and racial pulls.

I should like to see the authority of the supranational regime restricted altogether to the field of security. Whether this would be possible I am not sure. Experience may point to the desirability of adding some authority over economic matters, since under modern conditions these are capable of causing national upsets that have in them the seeds of violent conflict. But I should prefer to see the function of the organization altogether limited to the tasks of security. I also should like to see this regime established through the strengthening of the United Nations, so as not to sacrifice continuity in the search for peace.

I do not hide from myself the great difficulties of establishing a world government, either a beginning without Russia or one with Russia. I am aware of the risks. Since I should not wish it to be permissible for any country that has joined the supranational organization to secede, one of these risks is a possible civil war. But I also believe that world government is certain to come in time, and that the question is how much it is to be permitted to cost. It will come, I believe, even if there is another world war, though after such a war, if it is won, it would be world government established by the victor, resting on the victor's military power, and thus to be maintained permanently only through the permanent militarization of the human race.

But I also believe it can come through agreement and through the force of persuasion alone, hence at low cost. But if it is to come in this way it will not be enough to appeal to reason. One strength of the communist system of the East is that it has some of the character of a religion and inspires the emotions of a religion. Unless the cause of peace based on law gathers behind it the force and zeal of a religion, it hardly can hope to succeed. Those to whom the moral teaching of the human race is entrusted surely have a great duty and a great opportunity. The atomic scientists, I think, have become convinced that they cannot arouse the American people to the truths of the atomic era by logic alone. There must be added that deep power of emotion which is a basic ingredient of religion. It is to be hoped that not only the churches but the schools, the colleges, and the leading organs of opinion will acquit themselves well of their unique responsibility in this regard.

15

The War Is Won But Peace Is Not

PHYSICISTS FIND THEMSELVES in a position not unlike that of Alfred Nobel. Alfred Nobel invented the most powerful explosive ever known up to his time, a means of destruction par excellence. In order to atone for this, in order to relieve his human conscience he instituted his awards for the promotion of peace and for achievements of peace. Today, the physicists who participated in forging the most formidable and dangerous weapon of all times are harassed by an equal feeling of responsibility, not to say guilt. And we cannot desist from warning, and warning again, we cannot and should not slacken in our efforts to make the nations of the world, and especially their governments, aware of the unspeakable disaster they are certain to provoke unless they change their attitude toward each other and toward the task of shaping the future. We helped in creating this new weapon in order to prevent the enemies of mankind from achieving it ahead of us, which, given the mentality of the Nazis, would have meant inconceivable destruction and the enslavement of the rest of the world. We delivered this weapon into the hands of the American and the British people as trustees of the whole of mankind, as fighters for peace and liberty. But so far we fail to see any guarantee of peace, we do not see any guarantee of the freedoms that were promised to the nations in the Atlantic Charter. The war is won, but the peace is not. The great powers, united in fighting, are now divided over

the peace settlements. The world was promised freedom
from fear, but in fact fear has increased tremendously since
the termination of the war. The world was promised freedom from want, but large parts of the world are faced with
starvation while others are living in abundance. The nations
were promised liberation and justice. But we have witnessed,
and are witnessing even now, the sad spectacle of "liberating"
armies firing into populations who want their independence
and social equality, and supporting in those countries, by
force of arms, such parties and personalities as appear to be
most suited to serve vested interests. Territorial questions
and arguments of power, obsolete though they are, still prevail
over the essential demands of common welfare and justice.
Allow me to be more specific about just one case, which is
but a symptom of the general situation: the case of my own
people, the Jewish people.

As long as Nazi violence was unleashed only, or mainly,
against the Jews the rest of the world looked on passively, and
even treaties and agreements were made with the patently
criminal government of the Third Reich. Later, when Hitler
was on the point of taking over Rumania and Hungary, at the
time when Maidanek and Oswiecim were in Allied hands, and
the methods of the gas chambers were well known all over
the world, all attempts to rescue the Rumanian and Hungar-
ian Jews came to naught because the doors of Palestine were
closed to Jewish immigrants by the British government, and
no country could be found that would admit those forsaken
people. They were left to perish like their brothers and sisters
in the occupied countries.

We shall never forget the heroic efforts of the small countries, of the Scandinavian, the Dutch, the Swiss nations, and
of individuals in the occupied parts of Europe who did all

in their power to protect Jewish lives. We do not forget the humane attitude of the Soviet Union who was the only one among the big powers to open her doors to hundreds of thousands of Jews when the Nazi armies were advancing in Poland. But after all that has happened, and was not prevented from happening, how is it today? While in Europe territories are being distributed without any qualms about the wishes of the people concerned, the remainders of European Jewry, one fifth of its pre-war population, are again denied access to their haven in Palestine and left to hunger and cold and persisting hostility. There is no country, even today, that would be willing or able to offer them a place where they could live in peace and security. And the fact that many of them are still kept in the degrading conditions of concentration camps by the Allies gives sufficient evidence of the shamefulness and hopelessness of the situation. These people are forbidden to enter Palestine with reference to the principle of democracy, but actually the Western powers, in upholding the ban of the White Paper, are yielding to the threats and the external pressure of five vast and underpopulated Arab States. It is sheer irony when the British Foreign Minister tells the poor lot of European Jews they should remain in Europe because their genius is needed there, and, on the other hand, advises them not to try to get at the head of the queue lest they might incur new hatred and persecution. Well, I am afraid, they cannot help it; with their six million dead they have been pushed at the head of the queue, of the queue of Nazi victims, much against their will.

The picture of our postwar world is not bright. As far as we, the physicists, are concerned, we are no politicians and it has never been our wish to meddle in politics. But we know a few things that the politicians do not know. And we feel the duty to speak up and to remind those responsible that there is no escape into easy comforts, there is no distance ahead for proceeding little by little and delaying the necessary changes into an indefinite future, there is no time left for petty bargaining. The situation calls for a courageous effort, for a radical change in our whole attitude, in the entire political concept. May the spirit that prompted Alfred Nobel to create his great institution, the spirit of trust and confidence, of generosity and brotherhood among men, prevail in the minds of those upon whose decisions our destiny rests. Otherwise human civilization will be doomed.

16

The Menace of Mass Destruction

EVERYONE IS AWARE OF the difficult and menacing situation in which human society—shrunk into one community with a common fate—finds itself, but only a few act accordingly. Most people go on living their everyday life: half frightened, half indifferent, they behold the ghostly tragi-comedy that is being performed on the international stage before the eyes and ears of the world. But on that stage, on which the actors under the floodlights play their ordained parts, our fate of tomorrow, life or death of the nations, is being decided.

It would be different if the problem were not one of things made by Man himself, such as the atomic bomb and other means of mass destruction equally menacing all peoples. It would be different, for instance, if an epidemic of bubonic plague were threatening the entire world. In such a case conscientious and expert persons would be brought together and they would work out an intelligent plan to combat the plague. After having reached agreement upon the right ways and means, they would submit their plan to the governments. Those would hardly raise serious objections but rather agree speedily on the measures to be taken. They certainly would never think of trying to handle the matter in such a way that their own nation would be spared whereas the next one would be decimated.

But could not our situation be compared to one of a menacing epidemic? People are unable to view this situation in its true light, for their eyes are blinded by passion. General fear and anxiety create hatred and aggressiveness. The adaptation to warlike aims and activities has corrupted the mentality of man; as a result, intelligent, objective and humane thinking has hardly any effect and is even suspected and persecuted as unpatriotic.

There are, no doubt, in the opposite camps enough people of sound judgment and sense of justice who would be capable and eager to work out together a solution for the factual difficulties. But the efforts of such people are hampered by the fact that it is made impossible for them to come together for informal discussions. I am thinking of persons who are accustomed to the objective approach to a problem and who will not be confused by exaggerated nationalism or other passions. This forced separation of the people of both camps I consider one of the major obstacles to the achievement of an acceptable solution of the burning problem of international security.

As long as contact between the two camps is limited to the official negotiations I can see little prospect for an intelligent agreement being reached, especially since considerations of national prestige as well as the attempt to talk out of the window for the benefit of the masses are bound to make reasonable progress almost impossible. What one party suggests officially is for that reason alone suspected and even made unacceptable to the other. Also behind all official negotiations stands—though veiled—the threat of naked power. The official method can lead to success only after spade-work of an informal nature has prepared the ground; the conviction that a mutually satisfactory solution can be reached must be gained first; then the actual negotiations can get under way with a fair promise of success.

We scientists believe that what we and our fellow-men do or fail to do within the next few years will determine the fate of our civilization. And we consider it our task untiringly to explain this truth, to help people realize all that is at stake, and to work, not for appeasement, but for understanding and ultimate agreement between peoples and nations of different views.

17

The Schools and the Problem of Peace

———————

BY VIRTUE OF its geographic situation the United States is in the fortunate position of being able to teach a rational pacifism in its schools, without having to fear for its security. For there is no serious danger of a military attack from the outside, and as a result no compulsion to educate youth in a military spirit. On the other hand there is the danger of treating this problem purely from the emotional point of view. Yet little is gained by mere wishful thinking, without a clear grasp of the essential difficulties of the problem.

In the first place it ought to be made clear to youth that the United States may be at any time drawn into military involvements, even though a direct attack on the country need scarcely be feared. Mere reference to America's participation in the last World War is sufficient proof of this. Even Americans can hope for true security against being drawn into military involvements only from a satisfactory solution of the problem of peace in general. It is necessary to warn against the view that political isolation of the United States from the outside can result in adequate security for Americans. Instead a serious interest in an international solution of the problem of peace must be awakened among young people. In particular must youth be given a clear understanding of the grave responsibility which American politicians have assumed by failing to support Wilson's grandly conceived plans after peace was concluded, thus impairing the effectiveness of the League of Nations.

It must be pointed out that the mere demand for disarmament is futile, so long as there are great nations who are prepared to attain their future position in the world by means of military expansion. The reasonableness of the position represented by France, for example—namely that the security of the individual countries must be insured by international institutions—must be set forth. To achieve such security international treaties for common defense against those who break the peace are necessary but not sufficient. Instead, military defense resources must become internationalized by amalgamation and exchange of forces on a grand scale to such an extent that the military forces stationed in any one country cannot possibly be used exclusively for the purpose of pursuing the goals of that country.

To prepare the nations for such effective insurance of the peace, this vital problem should be clearly and sharply brought to the attention of young people. The spirit of international solidarity too should be strengthened and national chauvinism combatted as a harmful force impeding progress.

Schools ought to be intent on presenting history from the point of view of progress and the growth of human civilization, rather than using it as a means for fostering in the minds of the growing generation the ideals of outward power and military successes. In my opinion the use of H. G. Wells' *World History* should be highly recommended from this aspect.

It is of indirect yet nevertheless considerable importance, finally, that in the teaching of geography and history a sympathetic understanding be fostered for the characteristics of the different peoples of the world, especially for those whom we are in the habit of describing as "primitive."

18

On Military Service

I STAND FIRMLY by the principle that a real solution of the problem of pacifism can be achieved only by the organization of a supranational court of arbitration, which, differing from the present League of Nations in Geneva, would have at its disposal the means of enforcing its decisions. In short, an international court of justice with a permanent military establishment, or better, police force. An excellent expression of this conviction of mine is contained in Lord Davies' book, *Force* (London, Ernst Benn, Ltd., 1934), the reading of which I strongly recommend to everyone who is seriously concerned with this fundamental problem of mankind.

Taking as starting point this fundamental conviction, I stand for every measure which appears to me capable of bringing mankind nearer to this goal. Up to a few years ago, the refusal to bear arms by courageous and self-sacrificing persons was such a measure; it is no longer—especially in Europe—a means to be recommended. When the great Powers had nearly equally democratic governments, and when none of these Powers founded its future plans on military aggression, the refusal to do military service on the part of a fairly large number of citizens might have induced the governments of these Powers to look favorably on international legal arbitration. Moreover, such refusals were apt to educate public opinion to real pacifism. The public came to consider as oppression any pressure brought by the State upon its citizens to force them to fulfill their military obligations, besides considering such pressure unethical from the moral standpoint. Under these circumstances, such refusals worked for the highest good.

Today, however, we are brought face to face with the fact that powerful States make independent opinions in politics impossible for their citizens, and lead their own people into error through the systematic diffusion of false information. At the same time, these States become a menace to the rest of the world by creating military organizations which encompass their entire population. This false information is spread by a muzzled press, a centralized radio service, and school education ruled by an aggressive foreign policy. In States of that description, refusal to perform military service means martyrdom and death for those courageous enough to object. In those States in which citizens still cling to some of their political rights, refusal to do military service means weakening the power of resistance of the remaining sane portions of the civilized world.

Because of this, no reasonable human being would today favor the refusal to do military service, at least not in Europe, which is at present particularly beset with dangers.

I do not believe that under present circumstances passive resistance is an effective method, even if carried out in the most heroic manner. Other times, other means, even if the final aim remains the same.

The confirmed pacifist must therefore at present seek a plan of action different from that of former, more peaceful times. He must try to work for this aim: That those States which favor peaceful progress may come as close together as possible in order to diminish the likelihood that the warlike programs of political adventurers whose States are founded on violence and brigandage will be realized. I have in mind, in the first place, well-considered and permanent concerted action on the part of the United States and the British Empire, together with France and Russia when possible.

Perhaps the present danger will facilitate this rapprochement and thus bring about pacifistic solution of international problems. This would be the hopeful side to the present dark situation; here consistent action can contribute much toward influencing public opinion in the right direction.

19

Military Intrusion in Science

THE MILITARY MENTALITY

IT SEEMS TO ME that the decisive point in the situation lies in the fact that the problem before us cannot be viewed as an isolated one. First of all, one may pose the following question: From now on institutions for learning and research will more and more have to be supported by grants from the state, since, for various reasons, private sources will not suffice. Is it at all reasonable that the distribution of the funds raised for these purposes from the taxpayer should be entrusted to the military? To this question every prudent person will certainly answer: "No!" For it is evident that the difficult task of the most beneficent distribution should be placed in the hands of people whose training and life's work give proof that they know something about science and scholarship.

If reasonable people, nevertheless, favor military agencies for the distribution of a major part of the available funds, the reason for this lies in the fact that they subordinate cultural concerns to their general political outlook. We must then focus our attention on these practical political viewpoints, their origins and their implications. In doing so we shall soon recognize that the problem here under discussion is but one of many, and can only be fully estimated and properly adjudged when placed in a broader framework.

The tendencies we have mentioned are something new

for America. They arose when, under the influence of the two World Wars and the consequent concentration of all forces on a military goal, a predominantly military mentality developed, which with the almost sudden victory became even more accentuated. The characteristic feature of this mentality is that people place the importance of what Bertrand Russell so tellingly terms "naked power" far above all other factors which affect the relations between peoples. The Germans, misled by Bismarck's successes in particular, underwent just such a transformation of their mentality—in consequence of which they were entirely ruined in less than a hundred years.

I must frankly confess that the foreign policy of the United States since the termination of hostilities has reminded me, sometimes irresistibly, of the attitude of Germany under Kaiser Wilhelm II, and I know that, independent of me, this analogy has most painfully occurred to others as well. It is characteristic of the military mentality that non-human factors (atom bombs, strategic bases, weapons of all sorts, the possession of raw materials, etc.) are held essential, while the human being, his desires and thoughts—in short, the psychological factors—are considered as unimportant and secondary. Herein lies a certain resemblance to Marxism, at least insofar as its theoretical side alone is kept in view. The individual is degraded to a mere instrument; he becomes "human materiel." The normal ends of human aspiration vanish with such a viewpoint. Instead, the military mentality raises "naked power" as a goal in itself—one of the strangest illusions to which men can succumb.

In our time the military mentality is still more dangerous than formerly because the offensive weapons have become much more powerful than the defensive ones. Therefore it leads, by necessity, to preventive war. The general insecurity that goes hand in hand with this results in the sacrifice of the citizen's civil rights to the supposed welfare of the state. Political witch-hunting, controls of all sorts (e.g., control of teaching and research, of the press, and so forth) appear inevitable, and for this reason do not encounter that popular resistance, which, were it not for the military mentality, would provide a protection. A reappraisal of all values gradually takes place insofar as everything that does not clearly serve the Utopian ends is regarded and treated as inferior.

I see no other way out of prevailing conditions than a far-seeing, honest and courageous policy with the aim of establishing security on supranational foundations. Let us hope that men will be found, sufficient in number and moral force, to guide the nation on this path so long as a leading role is imposed on her by external circumstances. Then problems such as have been discussed here will cease to exist.

20

International Security

GEOGRAPHICALLY THE AMERICANS are without doubt in an especially favorable position, and menace to this country through military attack need not necessarily be given serious consideration. Nevertheless they manifest a real interest in the building up an international court of arbitration for the purpose of settling peaceably all international disputes or disagreements and with power to guarantee indemnities. The World War has shown the fate of the nations to be closely interwoven, and the world-wide economic crisis teaches us all the same.

Therefore it is essential that the American youth direct their energies to the end that the United States shall take active part in all efforts toward making international order a reality. It is obvious that the war and the post-war period have been a source of great concern to many Americans. It also follows that the continued policy of aloofness would not only injure all mankind, but harm the United States as well.

21

Isaac Newton

REASON, OF COURSE, is weak, when measured against its never-ending task. Weak, indeed, compared with the follies and passions of mankind, which, we must admit, almost entirely control our human destinies, in great things and small. Yet the works of the understanding outlast the noisy bustling generations and spread light and warmth across the centuries. Consoled by this thought let us turn, in these unquiet days, to the memory of Newton, who three hundred years ago was given to mankind.

To think of him is to think of his work. For such a man can be understood only by thinking of him as a scene on which the struggle for eternal truth took place. Long before Newton there had been virile minds who conceived that it ought to be possible, by purely logical deduction from simple physical hypotheses, to make cogent explanations of phenomena perceptible to the senses. But Newton was the first to succeed in finding a clearly formulated basis from which he could deduce a wide field of phenomena by means of mathematical thinking, logically, quantitatively and in harmony with experience. Indeed, he might well hope that the fundamental basis of his mechanics would come in time to furnish the key to the understanding of all phenomena. So thought his pupils—with more assurance than he himself—and so his successors, up till the end of the eighteenth century. How did this miracle come to birth in his brain? Forgive me, reader, the illogical

question. For if by reason we could deal with the problem
of the "how," then there could be no question of a miracle in
the proper sense of the word. It is the goal of every activity
of the intellect to convert a "miracle" into something which
it has grasped. If in this case the miracle permits itself to be
converted, our admiration for the mind of Newton becomes
only the greater thereby.

Galileo, by ingenious interpretation of the simplest facts
of experience, had established the proposition: a body upon
which no external force is at work permanently maintains
its original velocity (and direction); if it alters its velocity (or
the direction of its movement) the change must be referred
to an external cause.

To utilize this knowledge quantitatively the conceptions
velocity and rate of change of velocity—that is, acceleration in the case of any given motion of a body conceived as
dimensionless (material point)—must first be interpreted
with mathematical exactness. The task led Newton to invent
the basis of differential and integral calculus.

This in itself was a creative achievement of the first order.
But for Newton, as a physicist, it was simply the invention of
a new kind of conceptual language which he needed in order
to formulate the general laws of motion. For a given body
he had now to put forward the hypothesis that his precisely
formulated acceleration both in magnitude and direction was
proportional to the force directed upon it. The coefficient of
proportionality which characterizes the body with reference
to its power of acceleration completely describes the (dimensionless) body with reference to its mechanical quality; thus
was discovered the fundamental conception of mass.

All the foregoing might be described—though in
the extremely modest manner of speaking—as an exact

formulation of something the essence of which had already been recognized by Galileo. But it by no means succeeded in solving the main problem. In other words, the law of motion yields the movement of a body, only when the direction and magnitude of the force exerted upon it are known for all times. Thus the problem reduced itself to another problem: how to find out the operative forces. To a mind any less bold than Newton's it must have seemed hopeless, considering the immeasurable multifarity of the effects which the bodies of a universe seem to produce upon each other. Moreover, the bodies whose motions we perceive are by no means dimensionless points—that is to say, perceptible as material points. How was Newton to deal with such chaos?

If we push a cart moving without friction on a horizontal plane it follows that the force we exert upon it is given directly. That is the ideal case from which the law of motion is derived. That we are not here dealing with a dimensionless point appears unessential.

How does it stand then with a falling body in space? A freely falling body behaves almost as simply as the dimensionless point, if one regards its movement as a whole. It is accelerated downwards. The acceleration, according to Galileo, is independent of its nature and its velocity. The earth, of course, must be decisive for the existence of this acceleration. It seemed, then, that the earth by its mere presence exerted a force upon the body. The earth consists of many parts. The idea seemed inevitable that each of these parts affects the falling body and that all these effects are combined. There seems then to be a force which bodies by their very presence exert upon each other through space. These forces seem to be independent of velocities, dependent only upon the relative position and quantitative property of the various bodies exerting them. This quantitative property might be conditioned by its mass, for the mass seems to characterize the body from the mechanical point of view. This strange effect of things at a distance may be called gravitation.

Now to gain precise knowledge of this effect, one has only to find out how strong is the force exerted upon each other by two bodies of given mass from a given distance. As for their direction, it would probably be no other than the line connecting them. Finally then, what remains unknown is only the dependence of this force upon the distance between the two bodies. But this one cannot know a priori. Here, only experience could be of use.

Such experience, however, was available to Newton.

The acceleration of the moon was known from its orbit and could be compared with the acceleration of the freely falling body on the surface of the earth. Furthermore, the movements of the planets about the sun had been determined by Kepler with great exactness and comprehended in simple empirical laws. So it was possible to ascertain how the effects of gravitation coming from the earth and those coming from the sun depended on the factor of distance. Newton found that everything was explainable by a force which was inversely proportional to the square of the distance. And with that the goal was reached, the science of celestial mechanics was born, confirmed a thousand times over by Newton himself and those who came after him. But how about the rest of physics? Gravitation and the law of motion could not explain everything. What determined the equilibrium of the parts of a solid body? How was light to be explained, how electrical phenomena? By introducing material points and forces of various kinds acting at a distance, everything seemed in a fair way to be derivable from the law of motion.

That hope has not been fulfilled, and no one any longer believes in the solution of all our problems on this basis. Nevertheless, the thinking of physicists today is conditioned to a high degree by Newton's fundamental conceptions. It has so far not been possible to substitute for the Newtonian unified conception of the universe a similarly unified comprehensive conception. But what we have gained up till now would have been impossible without Newton's clear system.

From observation of the stars have chiefly come the intellectual tools indispensable to the development of modern technique. For the abuse of the latter in our time creative intellects like Newton's are as little responsible as the stars themselves, contemplating which their thoughts took wing. It is necessary to say this, because in our time esteem for intellectual values for their own sake is no longer so lively as it was in the centuries of the intellectual renascence.

22

Johannes Kepler

IN KEPLER'S LETTERS we find ourselves confronted with
a sensitive personality, passionately devoted to the quest for
deeper insight into the character of natural processes—a
man who reached the exalted goal he set himself in spite of
all internal and external difficulties. Kepler's life was devoted
to the solution of a dual problem. The sun and the planets
change their apparent position with reference to their background of fixed stars in a complex manner open to immediate observation. In other words, all the observations and
records compiled with such care dealt not actually with the
movements of the planets in space but with temporal shifts
undergone by the direction earth-planet in the course of
time.

Once Copernicus had convinced the small group capable
of grasping it that in this process the sun must be regarded
as being at rest, with the planets, including the earth, revolv-
ing about the sun, the first great problem proved to be this:
to determine the true motions of the planets, including the
earth, as they might be visible to an observer on the nearest fixed star who was equipped with a perfect stereoscopic
double-telescope. This was Kepler's first great problem. The
second problem was embodied in this question: What are the
mathematical laws under which these motions proceed? It is
plain that the solution of the second problem, if at all within
reach of the human mind, was predicated on the solution of

the first. Before a theory explaining a certain process can be tested, that process must first be known.

Kepler's solution of the first problem is based on a truly inspired notion that made possible the determination of the true orbit of the earth. To construct that orbit, a second fixed point in planetary space, in addition to the sun, is required. When such a second point is available, it and the sun may both be used as points of reference for angular measurements, and the earth's true orbit can be determined by the same methods of triangulation that customarily serve in surveying and cartography.

But where was such a second fixed point to be found, since all visible objects, except the sun, themselves execute motions that are not known in detail? This was Kepler's answer: The apparent motions of the planet Mars are known with great accuracy, including the time of its revolution about the sun (the "Martian year"). It is probable that at the end of each Martian year Mars is at the same spot in (planetary) space. If we limit ourselves for the time being to these points in time, then the planet Mars represents for them a fixed point in planetary space, a point that may be used in triangulation.

Employing this principle, Kepler first of all determined the true motion of the earth in planetary space. Since the earth itself may be used as a point for triangulation at any time, he was also able to determine the true motions of the other planets from his observations.

This is how Kepler gained the basis for formulating the three fundamental laws with which his name will remain associated for all time to come. Today, after the fact, no one can fully appreciate how much ingenuity, how much hard and tireless work was required to discover these laws and ascertain them with such precision.

The reader ought to know this as he learns from the letters under what hardships Kepler accomplished this gigantic work. He refused to be paralyzed or discouraged either by poverty or by the lack of comprehension among those of his contemporaries who had the power to shape his life and work. Yet he was dealing with a subject that offered immediate danger to him who professed the truth. But Kepler was one of the few who are simply incapable of doing anything but stand up openly for their convictions in every field. At the same time he was not one who took undiluted pleasure in personal controversy, as was plainly the case with Galileo, whose inspired barbs delight the informed reader even today. Kepler was a devout Protestant, but he made no secret of the fact that he did not approve of all decisions by the Church. He was, accordingly, regarded as a kind of moderate heretic and treated as such.

This brings me to the inner difficulties Kepler had to overcome—difficulties at which I have already hinted. They are not as readily perceived as the outward difficulties. Kepler's lifework was possible only once he succeeded in freeing himself to a great extent of the intellectual traditions into which he was born. This meant not merely the religious tradition, based on the authority of the Church, but general concepts on the nature and limitations of action within the universe and the human sphere, as well as notions of the relative importance of thought and experience in science.

He had to rid himself of the animist approach in research, a mode of thought oriented toward ulterior ends. He first had to recognize that even the most lucidly logical mathematical theory was of itself no guarantee of truth, becoming meaningless unless it was checked against the most exacting observations in natural science. But for this philosophical orientation

Kepler's work would not have been possible. He does not speak of it, but the inner struggle is reflected in his letters. Let the reader watch out for remarks concerning astrology. They show that the vanquished inner foe had been rendered harmless, even though he was not yet altogether dead.

23

Marie Curie in Memoriam

AT A TIME WHEN a towering personality like Mme. Curie has come to the end of her life, let us not merely rest content with recalling what she has given to mankind in the fruits of her work. It. is the moral qualities of its leading personalities that are perhaps of even greater significance for a generation and for the course of history than purely intellectual accomplishments. Even these latter are, to a far greater degree than is commonly credited, dependent on the stature of character.

It was my good fortune to be linked with Mme. Curie through twenty years of sublime and unclouded friendship. I came to admire her human grandeur to an ever growing degree. Her strength, her purity of will, her austerity toward herself, her objectivity, her incorruptible judgment—all these were of a kind seldom found joined in a single individual. She felt herself at every moment to be a servant of society and her profound modesty never left any room for complacency. She was oppressed by an abiding sense for the asperities and inequities of society. This is what gave her that severe outward aspect, so easily misinterpreted by those who were not close to her—a curious severity unrelieved by any artistic strain. Once she had recognized a certain way as the right one, she pursued it without compromise and with extreme tenacity.

The greatest scientific deed of her life—proving the existence of radioactive elements and isolating them—owes its accomplishment not merely to bold intuition but to a devotion and tenacity in execution under the most extreme hardships imaginable, such as the history of experimental science has not often witnessed.

If but a small part of Mme. Curie's strength of character and devotion were alive in Europe's intellectuals, Europe would face a brighter future.

24

Max Planck in Memoriam

———————

A MAN TO whom it has been given to bless the world with a great creative idea has no need for the praise of posterity. His very achievement has already conferred a higher boon upon him.

Yet it is good—indeed, it is indispensable—that representatives of all who strive for truth and knowledge should be gathered here today from the four corners of the globe. They are here to bear witness that even in these times of ours, when political passion and brute force hang like swords over the anguished and fearful heads of men, the standard of our ideal search for truth is being held aloft undimmed. This ideal, a bond forever uniting scientists of all times and in all places, was embodied with rare completeness in Max Planck.

Even the Greeks had already conceived the atomistic nature of matter and the concept was raised to a high degree of probability by the scientists of the nineteenth century. But it was Planck's law of radiation that yielded the first exact determination—independent of other assumptions—of the absolute magnitudes of atoms. More than that, he showed convincingly that in addition to the atomistic structure of matter there is a kind of atomistic structure to energy, governed by the universal constant h, which was introduced by Planck.

This discovery became the basis of all twentieth-century research in physics and has almost entirely conditioned its development ever since. Without this discovery it would not have been possible to establish a workable theory of molecules and atoms and the energy processes that govern their transformations. Moreover, it has shattered the whole framework of classical mechanics and electrodynamics and set science a fresh task: that of finding a new conceptual basis for all physics. Despite remarkable partial gains, the problem is still far from a satisfactory solution.

In paying homage to this man the American National Academy of Sciences expresses its hope that free research, for the sake of pure knowledge, may remain unhampered and unimpaired.

25

Paul Langevin in Memoriam

THE NEWS OF Paul Langevin's death dealt me a greater blow than most of the events of these fateful years, so fraught with disappointment. Why should this have been the case? Was his not a long life, crowded with fruitful creative work—the life of a man in harmony with himself? Was he not widely revered for his keen insight into intellectual problems, universally beloved for his devotion to every good cause, for his understanding kindness toward all creatures? Is there not a certain satisfaction in the fact that natural limits are set to the life of the individual, so that at its conclusion it may appear as a work of art?

The sorrow brought on by Paul Langevin's passing has been so particularly poignant because it has given me a feeling of being left utterly alone and desolate. There are so very few in any one generation, in whom clear insight into the nature of things is joined with an intense feeling for the challenge of true humanity and the capacity for militant action. When such a man departs, he leaves a gap that seems unbearable to his survivors.

Langevin was endowed with unusual clarity and agility in scientific thought, together with a sure intuitive vision for the essential points. It was as a result of these qualities that his lectures exerted a crucial influence on more than one generation of French theoretical physicists. But Langevin also knew a great deal about experimental technique and his

criticism and constructive suggestions always carried a fruitful effect. His own original researches, moreover, decisively
influenced the development of science, mainly in the fields
of magnetism and ion theory. Yet the burden of responsibility which he was always ready to assume circumscribed his
own research work, so that the fruits of his labors emerge in
the publications of other scientists to a greater extent than
in his own.

It appears to me as a foregone conclusion that he would
have developed the Special Theory of Relativity, had that
not been done elsewhere; for he had clearly perceived its
essential aspects. Another admirable thing is that he fully
appreciated the significance of De Broglie's ideas—from
which Schrödinger subsequently developed the methods
of wave mechanics—even before these ideas had become
consolidated into a consistent theory. I vividly recall the
pleasure and warmth with which he told me about it—and
I also remember that I followed his remarks but hesitantly
and doubtfully.

All his life Langevin suffered from an awareness of the
deficiencies and inequities of our social and economic institutions. Yet he believed firmly in the power of reason and
knowledge. So pure in heart was he that he was convinced
all men should be ready for complete personal renunciation,
once they had seen the light of reason and justice. Reason
was his creed—a creed that was to bring not only light but
also salvation. His desire to promote the happier life for all
men was perhaps even stronger than his craving for pure
intellectual enlightenment. Thus it was that he devoted much
of his time and vital energy to political enlightenment. No
one who appealed to his social conscience ever went away
from him empty-handed. Thus it was too that the very moral

grandeur of his personality earned him the bitter enmity of many of the more humdrum intellectuals. He in turn understood them all and in his kindness never harbored resentment against anyone.

I can only give expression to my gratitude for having personally known this man of purity and illumination.

26

Walther Nernst in Memoriam

—————

WALTHER NERNST, WHO DIED recently, was one of the most characteristic and most interesting scholars with whom I have been closely connected during my life. He did not miss any of the conferences on physics in Berlin, and his brief remarks gave evidence of a truly amazing scientific instinct combined both with a sovereign knowledge of an enormous volume of factual materials, which was always at his command, and with a rare mastery of the experimental methods and tricks in which he excelled. Although sometimes good-naturedly smiling at his childlike vanity and self-complacency, we all had for him not only a sincere admiration, but also a personal affection. So long as his egocentric weakness did not enter the picture, he exhibited an objectivity very rarely found, an infallible sense for the essential, and a genuine passion for knowledge of the deep interrelations of nature. But for such a passion his singularly creative productivity and his important influence on the scientific life of the first third of this century would not have been possible. He ascended from Arrhenius, Ostwald and Van't Hoff, as the last of a dynasty which based their investigations on thermodynamics, osmotic pressure and ionic theory. Up to 1905 his work was essentially restricted to that range of ideas. His theoretical equipment was somewhat elementary, but he mastered it with a rare ingenuity. I refer, for instance, to the theory of electromotive powers in solutions of locally variable concentration, the theory of diminution of the solubility by adding a dissolved substance. During this period he invented the witty null-method of determining the dielectric constant of electrically conducting bodies by means of Wheatstone's Bridge (alternating current, telephone as indicator, compensating capacity in comparison-bridge branches).

This first productive period is largely concerned with improving the methodology and completing the exploration of a field the principles of which had already been known before Nernst. This work led him gradually to a general problem which is characterized by the question: Is it possible to compute from the known energy of the conditions of a system, the useful work which is to be gained by its transition from one state into another? Nernst realized that a theoretical determination of the transition work A from the energy-difference U by means of equations of thermodynamics alone is not possible. There could be inferred from thermodynamics that, at absolute zero, the temperature of the quantities A and U must be equal. But one could not derive A from U for any arbitrary temperatures, even if the energy-values or differences in U were known for all conditions. This computation was not possible until there was introduced, with regard to the reaction of these quantities under low temperatures, an assumption which appeared obvious because of its simplicity. This assumption is simply that A becomes temperature-independent under low temperatures. The introduction of this assumption as a hypothesis (third main principle of the theory of heat) is Nernst's greatest contribution to theoretical science. Planck found later a solution which is theoretically more satisfactory; namely, the entropy disappears at absolute zero temperature.

From the standpoint of the older ideas on heat, this third

main principle required very strange reactions of bodies under low temperatures. To pass upon the correctness of this principle, the methods of calorimetry under low temperatures had to be greatly improved. The calorimetry of high temperatures also owes to Nernst considerable progress. Through all these investigations, as well as through many stimulating suggestions with which his untiring inventive genius supplied experimenters in his field, he promoted the research work of his generation most effectively. The beginnings of the quantum theory were assisted by the important results of those caloric investigations, and this especially before Bohr's theory of the atom made spectroscopy the most important experimental field. Nernst's standard work, "Theoretical Chemistry," offers, not only to the student but also to the scholar, an abundance of stimulating ideas; it is theoretically elementary, but clever, vivid and full of intimations of manifold interrelations. It truly reflects his intellectual characteristics.

Nernst was not a one-sided scholar. His sound common sense engaged successfully in all fields of practical life, and every conversation with him brought something interesting to light. What distinguished him from almost all his fellow-countrymen was his remarkable freedom from prejudices. He was neither a nationalist nor a militarist. He judged things and people almost exclusively by their direct success, not by a social or ethical ideal. This was a consequence of his freedom from prejudices. At the same time he was interested in literature and had such a sense of humor as is very seldom found with men who carry so heavy a load of work. He was an original personality; I have never met any one who resembled him in any essential way.

27

Paul Ehrenfest in Memoriam

──────────

IT HAPPENS SO OFTEN nowadays that men of high qualities depart this life of their own free will that we no longer
feel such a conclusion to be unusual. Yet the decision to
take leave generally stems from an incapacity—or at least
an unwillingness—to resign oneself to new and more difficult outward conditions of life. To refuse to live out one's
natural life because of inner conflicts that are felt to be
intolerable—that is even today in persons of sound mind
a rare occurrence, possible only in the case of the noblest
and morally most exalted personalities. It is to such a tragic
inner conflict that our friend Paul Ehrenfest has succumbed.
Those who knew him well, as was vouchsafed to me, know
that this unblemished personality in the main fell victim to
a conflict of conscience that in some form or other is spared
no university teacher who has passed, say, his fiftieth year.

I came to know him twenty-two years ago. He visited
me in Prague, coming straight from Russia where he as a Jew
was debarred from teaching at institutions of higher learning.
He was looking for a sphere of work in central or western
Europe. But we talked little of that, for it was the state of science at the time that took up almost all of our interest. Both
of us realized that classical mechanics and the theory of the
electric field had failed in the face of the phenomena of heat
radiation and molecular processes (the statistical theory
of heat), but there seemed to be no feasible way out of this

dilemma. The logical gap in Planck's Theory of Radiation—which we both, nevertheless, greatly admired—was apparent to us. We also discussed the Theory of Relativity, to which he responded with a certain skepticism but with the critical judgment peculiar to him. Within a few hours we were true friends—as though our dreams and aspirations were meant for each other. We remained joined in close friendship until he departed this life.

His stature lay in his unusually well developed faculty to grasp the essence of a theoretical notion, to strip a theory of its mathematical accouterments until the simple basic idea emerged with clarity. This capacity made him a peerless teacher. It was on its account that he was invited to scientific congresses; for he always brought clarity and acuteness into any discussion. He fought against fuzziness and circumlocution, when necessary employing his sharp wit and even downright discourtesy. Some of his utterances could have been interpreted almost as arrogant, yet his tragedy lay precisely in an almost morbid lack of self-confidence. He suffered incessantly from the fact that his critical faculties transcended his constructive capacities. In a manner of speaking, his critical sense robbed him of his love for the offspring of his own mind even before they were born.

Shortly after our first encounter there occurred the great turning-point in Ehrenfest's outward career. Our revered master, Lorentz, anxious to retire from regular university teaching, had recognized Ehrenfest for the inspired teacher that he was and recommended him as his successor. A marvelous sphere of activity opened up to the still youthful man. He was not merely the best teacher in our profession whom I have ever known; he was also passionately preoccupied with the development and destiny of men, especially his students.

To understand others, to gain their friendship and trust, to aid anyone embroiled in outer or inner struggles, to encourage youthful talent—all this was his real element, almost more than immersion in scientific problems. His students and colleagues in Leyden loved and esteemed him. They knew his utter devotion, his nature so wholly attuned to service and help. Should he not have been a happy man?

In truth he felt unhappier than anyone else who was close to me. The reason was that he did not feel equal to the lofty task that confronted him. Of what use was it that everyone held him in esteem? His sense of inadequacy, objectively unjustified, plagued him incessantly, often robbing him of the peace of mind necessary for tranquil research. So greatly did he suffer that he was compelled to seek solace in distraction. His frequent aimless travels, his preoccupation with the radio, and many other features of his restless life stemmed not from a need for composure and harmless hobbies but rather from a curious urge for escape caused by the psychic conflict at which I have hinted.

In the last few years this situation was aggravated by the strangely turbulent development which theoretical physics has recently undergone. To learn and to teach things that one cannot fully accept in one's heart is always a difficult matter, doubly difficult for a mind of fanatical honesty, a mind to which clarity means everything. Added to this was the increasing difficulty of adaptation to new thoughts which always confronts the man past fifty. I do not know how many readers of these lines will be capable of fully grasping that tragedy. Yet it was this that primarily occasioned his escape from life.

It seems to me that the tendency toward exaggerated self-criticism is associated with experiences in boyhood.

Humiliation and mental oppression by ignorant and selfish teachers wreak havoc in the youthful mind that can never be undone and often exert a baleful influence in later life. The intensity of such experiences in Ehrenfest's case may be judged from the fact that he refused to entrust his dearly beloved children to any school.

His relations with his friends played a far greater role in Ehrenfest's life than is the case with most men. He was virtually dominated by his sympathies and also by antipathies based on moral judgments. The strongest relationship in his life was that to his wife and fellow worker, an unusually strong and steadfast personality and his intellectual equal. Perhaps her mind was not quite as agile, versatile, and sensitive as his own, but her poise, her independence of others, her steadfastness in the face of all hardships, her integrity in thought, feeling, and action—all these were a blessing to him and he repaid her with a veneration and love such as I have not often witnessed in my life. A fateful partial estrangement from her was a frightful experience for him, one with which his already wounded soul was unable to cope.

We whose lives have been enriched by the power and integrity of his spirit, the kindness and warmth of his rich mind, and not least his irrepressible humor and trenchant wit—we know how much his departure has impoverished us. He will live on in his students and in all whose aspirations were guided by his personality.

28

Mahatma Gandhi

———————

A LEADER OF his people, unsupported by any outward authority: a politician whose success rests not upon craft nor the mastery of technical devices, but simply on the convincing power of his personality; a victorious fighter who has always scorned the use of force; a man of wisdom and humility, armed with resolve and inflexible consistency, who has devoted all his strength to the up lifting of his people and the betterment of their lot; a man who has confronted the brutality of Europe with the dignity of the simple human being, and thus at all times risen superior.

Generations to come, it may be, will scarce believe that such a one as this ever in flesh and blood walked upon this earth.

29

Carl von Ossietzky

ONLY ONE WHO SPENT the years following the First World War in Germany can fully understand how hard a battle it was that a man like Ossietzky had to fight. He knew that the tradition of his countrymen, bent on violence and war, had not lost its power. He knew how difficult, thankless and dangerous a task it was, to preach sanity and justice to his countrymen who had been hardened by a rough fate and the demoralizing influence of a long war. In their blindness they repaid him in hatred, persecution and slow destruction; to heed him and to act accordingly would have meant their salvation and would have been a true relief for the whole world.

It will be to the eternal fame of the Nobel Foundation that it bestowed its high honor on this humble martyr, and that it is resolved to keep alive his memory and the memory of his work. It is also wholesome for mankind today, since the fatal illusion against which he fought has not been removed by the outcome of the last war. The abstention from the solution of human problems by brute force—is the task today as it was then.

30

Why Do They Hate the Jews?

I SHOULD LIKE to begin by telling you an ancient fable, with a few minor changes—a fable that will serve to throw into bold relief the mainsprings of political anti-Semitism:

The shepherd boy said to the horse: "You are the noblest beast that treads the earth. You deserve to live in untroubled bliss; and indeed your happiness would be complete were it not for the treacherous stag. But he practiced from youth to excel you in fleetness of foot. His faster pace allows him to reach the water holes before you do. He and his tribe drink up the water far and wide, while you and your foal are left to thirst. Stay with me! My wisdom and guidance shall deliver you and your kind from a dismal and ignominious state."

Blinded by envy and hatred of the stag, the horse agreed. He yielded to the shepherd lad's bridle. He lost his freedom and became the shepherd's slave.

The horse in this fable represents a people, and the shepherd lad a class or clique aspiring to absolute rule over the people; the stag, on the other hand, represents the Jews.

I can hear you say: "A most unlikely tale! No creature would be as foolish as the horse in your fable." But let us give it a little more thought. The horse had been suffering the pangs of thirst, and his vanity was often pricked when he saw the nimble stag outrunning him. You, who have known no such pain and vexation, may find it difficult to understand that hatred and blindness should have driven the horse to act with such ill-advised, gullible haste. The horse, however, fell an easy victim to temptation because his earlier tribulations had prepared him for such a blunder. For there is much truth in the saying that it is easy to give just and wise counsel—to others!—but hard to act justly and wisely for oneself. I say to you with full conviction: We all have often played the tragic role of the horse and we are in constant danger of yielding to temptation again.

The situation illustrated in this fable happens again and again in the life of individuals and nations. In brief, we may call it the process by which dislike and hatred of a given person or group are diverted to another person or group incapable of effective defense. But why did the role of the stag in the fable so often fall to the Jews? Why did the Jews so often happen to draw the hatred of the masses? Primarily because there are Jews among almost all nations and because they are everywhere too thinly scattered to defend themselves against violent attack.

A few examples from the recent past will prove the point: Toward the end of the nineteenth century the Russian people were chafing under the tyranny of their government. Stupid blunders in foreign policy further strained their temper until it reached the breaking point. In this extremity the rulers of Russia sought to divert unrest by inciting the masses to hatred and violence toward the Jews. These tactics were repeated after the Russian government had drowned the dangerous revolution of 1905 in blood—and this maneuver may well have helped to keep the hated regime in power until near the end of the World War.

When the Germans had lost the World War hatched by their ruling class, immediate attempts were made to blame the Jews, first for instigating the war and then for losing it. In the course of time, success attended these efforts. The hatred engendered against the Jews not only protected the privileged classes, but enabled a small, unscrupulous and insolent group to place the German people in a state of complete bondage.

The crimes with which the Jews have been charged in the course of history—crimes which were to justify the atrocities perpetrated against them—have changed in rapid succession. They were supposed to have poisoned wells. They were said to have murdered children for ritual purposes. They were falsely charged with a systematic attempt at the economic domination and exploitation of all mankind. Pseudo-scientific books were written to brand them an inferior, dangerous race. They were reputed to foment wars and revolutions for their own selfish purposes. They were presented at once as dangerous innovators and as enemies of true progress. They were charged with falsifying the culture of nations by penetrating the national life under the guise of becoming assimilated. In the same breath they were accused of being so stubbornly inflexible that it was impossible for them to fit into any society.

Almost beyond imagination were the charges brought against them, charges known to their instigators to be untrue all the while, but which time and again influenced the masses. In times of unrest and turmoil the masses are inclined to hatred and cruelty, whereas in times of peace these traits of human nature emerge but stealthily.

Up to this point I have spoken only of violence and oppression against the Jews—not of anti-Semitism itself as a psychological and social phenomenon existing even in times and circumstances when no special action against the Jews is under way. In this sense, one may speak of latent anti-Semitism. What is its basis? I believe that in a certain sense one may actually regard it as a normal manifestation in the life of a people.

The members of any group existing in a nation are more closely bound to one another than they are to the remaining population. Hence a nation will never be free of friction while such groups continue to be distinguishable. In my belief, uniformity in a population would not be desirable, even if it were attainable. Common convictions and aims, similar interests, will in every society produce groups that, in a certain sense, act as units. There will always be friction between such groups—the same sort of aversion and rivalry that exists between individuals.

The need for such groupings is perhaps most easily seen in the field of politics, in the formation of political parties. Without parties the political interests of the citizens of any state are bound to languish. There would be no forum for the free exchange of opinions. The individual would be isolated and unable to assert his convictions. Political convictions, moreover, ripen and grow only through mutual stimulation and criticism offered by individuals of similar disposition and purpose; and politics is no different from any other field of our cultural existence. Thus it is recognized, for example, that in times of intense religious fervor different sects are likely to spring up whose rivalry stimulates religious life in general. It is well known, on the other hand, that centralization—that is, elimination of independent groups—leads to one-sidedness and barrenness in science and art because such centralization checks and even suppresses any rivalry of opinions and research trends.

JUST WHAT IS A JEW?

The formation of groups has an invigorating effect in all spheres of human striving, perhaps mostly due to the struggle between the convictions and aims represented by the different groups. The Jews too form such a group with a definite character of its own, and anti-Semitism is nothing but the antagonistic attitude produced in the non-Jews by the Jewish group. This is a normal social reaction. But for the political abuse resulting from it, it might never have been designated by a special name.

What are the characteristics of the Jewish group? What, in the first place, is a Jew? There are no quick answers to this question. The most obvious answer would be the following: A Jew is a person professing the Jewish faith. The superficial character of this answer is easily recognized by means of a simple parallel. Let us ask the question: What is a snail? An answer similar in kind to the one given above might be: A snail is an animal inhabiting a snail shell. This answer is not altogether incorrect; nor, to be sure, is it exhaustive; for the snail shell happens to be but one of the material products of the snail. Similarly, the Jewish faith is but one of the characteristic products of the Jewish community. It is, furthermore, known that a snail can shed its shell without thereby ceasing to be a snail. The Jew who abandons his faith (in the formal sense of the word) is in a similar position. He remains a Jew.

Difficulties of this kind appear whenever one seeks to explain the essential character of a group.

The bond that has united the Jews for thousands of years and that unites them today is, above all, the democratic ideal of social justice, coupled with the ideal of mutual aid and tolerance among all men. Even the most ancient religious

scriptures of the Jews are steeped in these social ideals, which
have powerfully affected Christianity and Mohammedanism
and have had a benign influence upon the social structure of
a great part of mankind. The introduction of a weekly day of
rest should be remembered here—a profound blessing to all
mankind. Personalities such as Moses, Spinoza and Karl Marx,
dissimilar as they may be, all lived and sacrificed themselves
for the ideal of social justice; and it was the tradition of their
forefathers that led them on this thorny path. The unique
accomplishments of the Jews in the field of philanthropy
spring from the same source.

The second characteristic trait of Jewish tradition is
the high regard in which it holds every form of intellectual
aspiration and spiritual effort. I am convinced that this great
respect for intellectual striving is solely responsible for the
contributions that the Jews have made toward the progress
of knowledge, in the broadest sense of the term. In view of
their relatively small number and the considerable external
obstacles constantly placed in their way on all sides, the
extent of those contributions deserves the admiration of all
sincere men. I am convinced that this is not due to any special
wealth of endowment, but to the fact that the esteem in which
intellectual accomplishment is held among the Jews creates
an atmosphere particularly favorable to the development of
any talents that may exist. At the same time a strong critical
spirit prevents blind obeisance to any mortal authority.

I have confined myself here to these two traditional traits,
which seem to me the most basic. These standards and ideals
find expression in small things as in large. They are transmitted from parents to children; they color conversation and
judgment among friends; they fill the religious scriptures; and
they give to the community life of the group its characteristic

stamp. It is in these distinctive ideals that I see the essence of Jewish nature. That these ideals are but imperfectly realized in the group—in its actual everyday life—is only natural. However, if one seeks to give brief expression to the essential character of a group, the approach must always be by the way of the ideal.

WHERE OPPRESSION IS A STIMULUS

In the foregoing I have conceived of Judaism as a community of tradition. Both friend and foe, on the other hand, have often asserted that the Jews represent a race; that their characteristic behavior is the result of innate qualities transmitted by heredity from one generation to the next. This opinion gains weight from the fact that the Jews for thousands of years have predominantly married within their own group. Such a custom may indeed preserve a homogeneous race—if it existed originally; it cannot produce uniformity of the race—if there was originally a racial intermixture. The Jews, however, are beyond doubt a mixed race, just as are all other groups of our civilization. Sincere anthropologists are agreed on this point; assertions to the contrary all belong to the field of political propaganda and must be rated accordingly.

Perhaps even more than on its own tradition, the Jewish group has thrived on oppression and on the antagonism it has forever met in the world. Here undoubtedly lies one of the main reasons for its continued existence through so many thousands of years.

The Jewish group, which we have briefly characterized in the foregoing, embraces about sixteen million people—less than one per cent of mankind, or about half as many as the population of present-day Poland. Their significance as a political factor is negligible. They are scattered over almost the entire earth and are in no way organized as a whole—which means that they are incapable of concerted action of any kind.

Were anyone to form a picture of the Jews solely from the utterances of their enemies, he would have to reach the conclusion that they represent a world power. At first sight that seems downright absurd; and yet, in my view, there is a certain meaning behind it. The Jews as a group may be powerless, but the sum of the achievements of their individual members is everywhere considerable and telling, even though these achievements were made in the face of obstacles. The forces dormant in the individual are mobilized, and the individual himself is stimulated to self-sacrificing effort, by the spirit that is alive in the group.

Hence the hatred of the Jews by those who have reason to shun popular enlightenment. More than anything else in the world, they fear the influence of men of intellectual independence. I see in this the essential cause for the savage hatred of Jews raging in present-day Germany. To the Nazi group the Jews are not merely a means for turning the resentment of the people away from themselves, the oppressors; they see the Jews as a nonassimilable element that cannot be driven into uncritical acceptance of dogma, and that, therefore—as long as it exists at all—threatens their authority because of its insistence on popular enlightenment of the masses.

Proof that this conception goes to the heart of the matter is convincingly furnished by the solemn ceremony of the burning of the books staged by the Nazi regime shortly after its seizure of power. This act, senseless from a political point of view, can only be understood as a spontaneous emotional outburst. For that reason it seems to me more revealing than many acts of greater purpose and practical importance.

In the field of politics and social science there has grown up a justified distrust of generalizations pushed too far. When thought is too greatly dominated by such generalizations, misinterpretations of specific sequences of cause and effect readily occur, doing injustice to the actual multiplicity of events. Abandonment of generalization, on the other hand, means to relinquish understanding altogether. For that reason I believe one may and must risk generalization, as long as one remains aware of its uncertainty. It is in this spirit that I wish to present in all modesty my conception of anti-Semitism, considered from a general point of view.

In political life I see two opposed tendencies at work, locked in constant struggle with each other. The first, optimistic, trend proceeds from the belief that the free unfolding of the productive forces of individuals and groups essentially leads to a satisfactory state of society. It recognizes the need for a central power, placed above groups and individuals, but concedes to such power only organizational and regulatory functions. The second, pessimistic, trend assumes that free interplay of individuals and groups leads to the destruction of society; it thus seeks to base society exclusively upon authority, blind obedience and coercion. Actually this trend is pessimistic only to a limited extent: for it is optimistic in regard to those who are, and desire to be, the bearers of power and authority. The adherents of this second trend are the enemies of the free groups and of education for independent thought. They are, moreover, the carriers of political anti-Semitism.

Here in America all pay lip service to the first, optimistic, tendency. Nevertheless, the second group is strongly represented. It appears on the scene everywhere, though for the most part it hides its true nature. Its aim is political and spiritual dominion over the people by a minority, by the circuitous route of control over the means of production. Its proponents have already tried to utilize the weapon of anti-Semitism as well as of hostility to various other groups. They will repeat the attempt in times to come. So far all such tendencies have failed because of the people's sound political instinct.

And so it will remain in the future, if we cling to the rule: Beware of flatterers, especially when they come preaching hatred.

31

The Dispersal of European Jewry

THE HISTORY OF the persecutions which the Jewish people have had to suffer is almost inconceivably long. Yet the war that is being waged against us in Central Europe today* falls into a special category of its own. In the past we were persecuted despite the fact that we were the people of the Bible; today, however, it is just because we are the people of the Book that we are persecuted. The aim is to exterminate not only ourselves but to destroy, together with us, that spirit expressed in the Bible and in Christianity which made possible the rise of civilization in Central and Northern Europe. If this aim is achieved Europe will become a barren waste. For human community life cannot long endure on a basis of crude force, brutality, terror, and hate.

Only understanding for our neighbors, justice in our dealings, and willingness to help our fellow men can give human society permanence and assure security for the individual. Neither intelligence nor inventions nor institutions can serve as substitutes for these most vital parts of education.

Many Jewish communities have been uprooted in the wake of the present upheaval in Europe. Hundreds of thousands of men, women, and children have been driven from their homes and made to wander in despair over the highways of the world. The tragedy of the Jewish people today

---

* Spoken in 1939.

is a tragedy which reflects a challenge to the fundamental structure of modern civilization.

One of the most tragic aspects of the oppression of Jews and other groups has been the creation of a refugee class. Many distinguished men in science, art, and literature have been driven from the lands which they enriched with their talents. In a period of economic decline these exiles have within them the possibilities for reviving economic and cultural effort; many of these refugees are highly skilled experts in industry and science. They have a valuable contribution to make to the progress of the world. They are in a position to repay hospitality with new economic development and the opening up of new opportunities of employment for native populations. I am told that in England the admission of refugees was directly responsible for giving jobs to 15,000 unemployed.

As one of the former citizens of Germany who have been fortunate enough to leave that country, I know I can speak for my fellow refugees, both here and in other countries, when I give thanks to the democracies of the world for the splendid manner in which they have received us. We, all of us, owe a debt of gratitude to our new countries, and each and every one of us is doing the utmost to show our gratitude by the quality of our contributions to the economic, social, and cultural work of the countries in which we reside.

It is, however, a source of gravest concern that the ranks of the refugees are being constantly increased. The developments of the past week have added several hundred thousand potential refugees from Czechoslovakia. Again we are confronted with a major tragedy for a Jewish community which had a noble tradition of democracy and communal service. The power of resistance which has enabled the Jewish people to survive for thousands of years is a direct outgrowth of Jewish adherence to the Biblical doctrines on the relationships among men. In these years of affliction our readiness to help one another is being put to an especially severe test. Each of us must personally face this test, that we may stand it as well as our fathers did before us. We have no other means of self-defense than our solidarity and our knowledge that the cause for which we are suffering is a momentous and sacred cause.

32

Let's Not Forget

IF WE AS JEWS can learn anything from these politically sad times, it is the fact that destiny has bound us together, a fact which in times of quiet and security, we often so easily and gladly forget. We are accustomed to lay too much emphasis on the differences that divide the Jews of different lands and different religious views. And we forget often that it is the concern of every Jew, when anywhere the Jew is hated and treated unjustly, when politicians with flexible consciences set into motion against us the old prejudices, originally religious, in order to concoct political schemes at our expense. It concerns every one of us because such diseases and psychotic disturbances of the folk-soul are not estopped by oceans and national borders, but act precisely like economic crises and epidemics.

33

Unpublished Preface to a Blackbook

———

THIS BOOK IS a collection of documentary material on the systematic work of destruction by which the German Government murdered a great proportion of the Jewish people. Responsibility for the truth of the facts set forth is borne by the Jewish organizations that have joined to create the present work and present it to the public.

The purpose of this publication is manifest. It is to convince the reader that an international organization for safeguarding the sanctity of life can effectively fulfill its purpose only if it does not limit itself to protecting countries against military attack but also extends its protection to national minorities within the individual countries. For in the last reckoning it is the individual who must be protected against annihilation and inhuman treatment.

It is true that this goal can be attained only if the principle of non-intervention, which has played such a fateful role in the last decades, is cast overboard. Yet today no one can doubt the need for this far-reaching step any longer. For even those who envision only the attainment of protection against military attack from the outside must today realize that the disasters of war are preceded by certain internal developments in the various countries, and not merely by military and armaments preparations.

Not until the creation and maintenance of decent

conditions of life for all men are recognized and accepted as a common obligation of all men and all countries—not until then shall we, with a certain degree of justification, be able to speak of mankind as civilized.

Percentagewise the Jewish people have lost more than any other people affected by the disasters of recent years. If a truly just settlement is to be striven for, the Jewish people must be given special consideration in the organization of the peace. The fact that the Jews, in the formal political sense, cannot be regarded as a nation, insofar as they possess no country and no government, ought to be no impediment. For the Jews have been treated as a uniform group, as though they were a nation. Their status as a uniform political group is proved to be a fact by the behavior of their enemies. Hence in striving toward a stabilization of the international situation they should be considered as though they were a nation in the customary sense of the word.

Another factor must be emphasized in this connection. In parts of Europe Jewish life will probably be impossible for years to come. In decades of hard work and voluntary financial aid the Jews have restored the soil of Palestine to fertility. All these sacrifices were made because of trust in the officially sanctioned promise given by the governments in question after the last war, namely that the Jewish people were to be given a secure home in their ancient Palestinian country. To put it mildly, the fulfillment of this promise has been but hesitant and partial. Now that the Jews—especially the Jews in Palestine—have in this war too rendered a valuable contribution, the promise must be forcibly called to mind. The demand must be put forward that Palestine, within the limits of its economic capacity, be thrown open to Jewish immigration. If supranational institutions are to win that confidence that must form the most important buttress for their endurance, then it must be shown above all that those who, trusting to these institutions, have made the heaviest sacrifices are not defrauded.

34

The Goal of Human Existence

OUR AGE IS PROUD of the progress it has made in man's intellectual development. The search and striving for truth and knowledge is one of the highest of man's qualities—though often the pride is most loudly voiced by those who strive the least. And certainly we should take care not to make the intellect our god; it has, of course, powerful muscles, but no personality. It cannot lead, it can only serve; and it is not fastidious in its choice of a leader. This characteristic is reflected in the qualities of its priests, the intellectuals. The intellect has a sharp eye for methods and tools, but is blind to ends and values. So it is no wonder that this fatal blindness is handed on from old to young and today involves a whole generation.

Our Jewish forbears, the prophets and the old Chinese sages understood and proclaimed that the most important factor in giving shape to our human existence is the setting up and establishment of a goal; the goal being a community of free and happy human beings who by constant inward endeavor strive to liberate themselves from the inheritance of anti-social and destructive instincts. In this effort the intellect can be the most powerful aid. The fruits of intellectual effort, together with the striving itself, in cooperation with the creative activity of the artist, lend content and meaning to life.

But today the rude passions of man reign in our world, more unrestrained than ever before. Our Jewish people, a

139

small minority everywhere, with no means of defending themselves by force, are exposed to the crudest suffering, even to complete annihilation, to a far greater degree than any other people in the world. The hatred raging against us is grounded in the fact that we have upheld the ideal of harmonious partnership and given it expression in word and deed among the best of our people.

35

Our Debt to Zionism

RARELY SINCE THE CONQUEST of Jerusalem by Titus has the Jewish community experienced a period of greater oppression than prevails at the present time. In some respects, indeed, our own time is even more troubled, for man's possibilities of emigration are more limited today than they were then.

Yet we shall survive this period too, no matter how much sorrow, no matter how heavy a loss in life it may bring. A community like ours, which is a community purely by reason of tradition, can only be strengthened by pressure from without. For today every Jew feels that to be a Jew means to bear a serious responsibility not only to his own community, but also toward humanity. To be a Jew, after all, means first of all, to acknowledge and follow in practice those fundamentals in humaneness laid down in the Bible—fundamentals without which no sound and happy community of men can exist.

We meet today because of our concern for the development of Palestine. In this hour one thing, above all, must be emphasized: Judaism owes a great debt of gratitude to Zionism. The Zionist movement has revived among Jews the sense of community. It has performed productive work surpassing all the expectations any one could entertain. This productive work in Palestine, to which self-sacrificing Jews throughout the world have contributed, has saved a large number of our brethren from direst need. In particular, it has been possible to lead a not inconsiderable part of our youth toward a life of joyous and creative work.

Now the fateful disease of our time—exaggerated nationalism, borne up by blind hatred—has brought our work in Palestine to a most difficult stage. Fields cultivated by day must have armed protection at night against fanatical Arab outlaws. All economic life suffers from insecurity. The spirit of enterprise languishes and a certain measure of unemployment (modest when measured by American standards) has made its appearance.

The solidarity and confidence with which our brethren in Palestine face these difficulties deserve our admiration. Voluntary contributions by those still employed keep the unemployed above water. Spirits remain high, in the conviction that reason and calm will ultimately reassert themselves. Everyone knows that the riots are artificially fomented by those directly interested in embarrassing not only ourselves but especially England. Everyone knows that banditry would cease if foreign subsidies were withdrawn.

Our brethren in other countries, however, are in no way behind those in Palestine. They, too, will not lose heart but will resolutely and firmly stand behind the common work. This goes without saying.

Just one more personal word on the question of partition. I should much rather see reasonable agreement with the Arabs on the basis of living together in peace than the creation of a Jewish state. Apart from practical consideration, my awareness of the essential nature of Judaism resists the idea of a Jewish state with borders, an army, and a measure of temporal power no matter how modest. I am afraid of the inner damage Judaism will sustain—especially from the development of a narrow nationalism within our own ranks, against which we have already had to fight strongly, even without a Jewish state. We are no longer the Jews of the Maccabee period. A return to a nation in the political sense of the word would be equivalent to turning away from the spiritualization of our community which we owe to the genius of our prophets. If external necessity should after all compel us to assume this burden, let us bear it with tact and patience.

One more word on the present psychological attitude of the world at large, upon which our Jewish destiny also depends. Anti-Semitism has always been the cheapest means employed by selfish minorities for deceiving the people. A tyranny based on such deception and maintained by terror must inevitably perish from the poison it generates within itself. For the pressure of accumulated injustice strengthens those moral forces in man which lead to a liberation and purification of public life. May our community through its suffering and its work contribute toward the release of those liberating forces.

36

To the Heroes of the Battle
of the Warsaw Ghetto

THEY FOUGHT AND DIED as members of the Jewish nation, in the struggle against organized bands of German murderers. To us these sacrifices are a strengthening of the bond between us, the Jews of all the countries. We strive to be one in suffering and in the effort to achieve a better human society, that society which our prophets have so clearly and forcibly set before us as a goal.

The Germans as an entire people are responsible for these mass murders and must be punished as a people if there is justice in the world and if the consciousness of collective responsibility in the nations is not to perish from the earth entirely. Behind the Nazi party stands the German people, who elected Hitler after he had in his book and in his speeches made his shameful intentions clear beyond the possibility of misunderstanding. The Germans are the only people who have not made any serious attempt of counter-action leading to the protection of the innocently persecuted. When they are entirely defeated and begin to lament over their fate, we must not let ourselves be deceived again, but keep in mind that they deliberately used the humanity of others to make preparation for their last and most grievous crime against humanity.

37

Before the Monument to
the Martyred Jews of the Warsaw
Ghetto

<hr />

THE MONUMENT BEFORE WHICH you have gathered today was built to stand as a concrete symbol of our grief over the irreparable loss our martyred Jewish nation has suffered. It shall also serve as a reminder for us who have survived to remain loyal to our people and to the moral principles cherished by our fathers. Only through such loyalty may we hope to survive this age of moral decay.

The more cruel the wrong that men commit against an individual or a people, the deeper their hatred and contempt for their victim. Conceit and false pride on the part of a nation prevent the rise of remorse for its crime. Those who have had no part in the crime, however, have no sympathy for the sufferings of the innocent victims of persecution and no awareness of human solidarity. That is why the remnants of European Jewry are languishing in concentration camps and the sparsely populated lands of this earth close their gates against them. Even our right, so solemnly pledged, to a national homeland in Palestine is being betrayed. In this era of moral degradation in which we live the voice of justice no longer has any power over men.

Let us clearly recognize and never forget this: That mutual cooperation and the furtherance of living ties between the

Jews of all lands is our sole physical and moral protection in the present situation. But for the future our hope lies in overcoming the general moral abasement which today gravely menaces the very existence of mankind. Let us labor with all our powers, however feeble, to the end that mankind recover from its present moral degradation and gain a new vitality and a new strength in its striving for right and justice as well as for a harmonious society.

38

The Calling of the Jews

—————

THIS IS A TIME when there seems to be a particular need for men of philosophical persuasion—that is to say, friends of wisdom and truth—to join together. For while it is true that our time has accumulated more knowledge than any earlier age, that love of truth and insight which lent wings to the spirit of the Renaissance has grown cold, giving way to sober specialization rooted in the material spheres of society rather than in the spiritual. But groups such as this one are devoted solely to spiritual aims. In centuries past Judaism clung exclusively to its moral and spiritual tradition. Its teachers were its only leaders. But with adaptation to a larger social whole this spiritual orientation has receded into the background, though even today the Jewish people owe to it their apparently indestructible vigor. If we are to preserve that vigor for the benefit of mankind, we must hold to that spiritual orientation toward life.

The Dance about the Golden Calf was not merely a legendary episode in the history of our forefathers—an episode that seems to me in its simplicity more innocent than that total adherence to material and selfish goals threatening Judaism in our own days. At this time a union of those who rally to the spiritual heritage of our people has supreme justification. This is all the more true for a group that is free of all historical and national narrowness. We Jews should be and remain the carriers and patrons of spiritual values. But we should also always be aware of the fact that these spiritual values are and always have been the common goal of all mankind.

39

Moses Maimonides

———————

THERE IS SOMETHING SUBLIME in the spectacle of men joining together in a spirit of harmony to honor the memory of a man whose life and work lie seven centuries in the past. This feeling is accentuated all the more sharply at a time in which passion and strife tend more than usually to obscure the influence of reasoned thought and balanced justice. In the bustle of everyday life our view grows clouded with desire and passion, and the voice of reason and justice is almost inaudible in the hubbub of the struggle of all against all. But the ferment of those times long past has long since been stilled, and scarcely more is left of it than the memory of those few who exerted a crucial and fruitful influence on their contemporaries and thus on later generations as well. Such a man was Maimonides.

Once the Teutonic barbarians had destroyed Europe's ancient culture, a new and finer cultural life slowly began to flow from two sources that had somehow escaped being altogether buried in the general havoc—the Jewish Bible and Greek philosophy and art. The union of these two sources, so different one from the other, marks the beginning of our present cultural epoch, and from that union, directly or indirectly, has sprung all that makes up the true values of our present-day life.

Maimonides was one of those strong personalities who by their writings and their human endeavors helped to bring about that synthesis, thus paving the way for later develop-ments. Just how this happened will be related to us tonight by friends whose studies have come closer than I to the heart of Maimonides' lifework and the history of the European mind.

May this hour of grateful remembrance serve to strengthen within us the love and esteem in which we hold the treasures of our culture, gained in such bitter struggle. Our fight to preserve those treasures against the present pow-ers of darkness and barbarism cannot then but carry the day.

40

Stephen Wise

AMONG ALL THOSE whom I have personally met who have labored in the cause of justice and in the interest of the hard-pressed Jewish people, only a few were at all times selfless—but there was no one who gave his love and energy with such consuming devotion as Stephen Wise. All his life he has been a fighter for the cause of Zionism to which the memory of his ceaseless activities will be bound for ever. He has walked the thorny paths of the true prophet, at all times disdaining sordid compromise and never bending the knee to those in power. By relentlessly exposing the weakness and imperfections both in our own ranks and in the larger political arena of the non-Jewish world, he has made great and lasting contributions wherever he has gone. There are those who do not love him, but there is no one who has ever denied him recognition and respect, for everybody knows that behind the enormous labors of this man there has always been the passionate desire to make mankind better and happier.

41

To the University of Jerusalem

THE LITTLE THAT I could do, in a long life favored by external circumstances to deepen our physical knowledge, has brought me so much praise that for a long time I have felt rather more embarrassed than elated. But from you there comes a token of esteem that fills me with pure joy—joy about the great deeds that our Jewish people have accomplished within a few generations, under exceptionally difficult conditions, by itself alone, through boundless courage and immeasurable sacrifices. The University which twenty-seven years ago was nothing but a dream and a faint hope, this University is today a living thing, a home of free learning and teaching and happy brotherly work. There it is, on the soil that our people have liberated under great hardships; there it is, a spiritual center of a flourishing and buoyant community whose accomplishments have finally met with the universal recognition they deserved.

In this last period of the fulfilment of our dreams there was but one thing that weighed heavily upon me: the fact that we were compelled by the adversities of our situation to assert our rights through force of arms; it was the only way to avert complete annihilation. The wisdom and moderation the leaders of the new state have shown gives me confidence, however, that gradually relations will be established with the Arab people which are based on fruitful cooperation and mutual respect and trust. For this is the only means through which both peoples can attain true independence from the outside world.

42

The American Council for Judiasm

I AM VERY HAPPY indeed to hear that the platform for which the American Council for Judaism stands is meeting with strong opposition. This organization appears to me to be nothing more than a pitiable attempt to obtain favor and toleration from our enemies by betraying true Jewish ideals and by mimicking those who claim to stand for 100 per cent Americanism. I believe this method to be both undignified and ineffective. Our opponents are bound to view it with disdain and even with contempt, and in my opinion justly. He who is untrue to his own cause cannot command the respect of others. Apart from these considerations, the movement in question is a fairly exact copy of the Zentralverein Deutscher Staatsbürger Jüdischen Glaubens ("Central Association of German Citizens of Jewish Faith") of unhappy memory, which in the days of our crucial need showed itself utterly impotent and corroded the Jewish group by undermining that inner certitude by which alone our Jewish people could have overcome the trials of this difficult age.

43

The Jews of Israel

THERE IS NO PROBLEM of such overwhelming importance
to us Jews as consolidating that which has been accomplished
in Israel with amazing energy and an unequalled willingness
for sacrifice. May the joy and admiration that fill us when we
think of all that this small group of energetic and thoughtful
people has achieved give us the strength to accept the great
responsibility which the present situation has placed upon us.

When appraising the achievement, however, let us not
lose sight of the cause to be served by this achievement: rescue of our endangered brethren, dispersed in many lands, by
uniting them in Israel; creation of a community which con-
forms as closely as possible to the ethical ideals of our people
as they have been formed in the course of a long history.

One of these ideals is peace, based on understanding
and self-restraint, and not on violence. If we are imbued
with this ideal, our joy becomes somewhat mingled with
sadness, because our relations with the Arabs are far from
this ideal at the present time. It may well be that we would
have reached this ideal, had we been permitted to work out,
undisturbed by others, our relations with our neighbors, for
we want peace and we realize that our future development
depends on peace.

It was much less our own fault or that of our neighbors
than of the Mandatory Power, that we did not achieve an
undivided Palestine in which Jews and Arabs would live as

equals, free, in peace. If one nation dominates other nations, as was the case in the British Mandate over Palestine, she can hardly avoid following the notorious device of Divide et Impera. In plain language this means: create discord among the governed people so they will not unite in order to shake off the yoke imposed upon them. Well, the yoke has been removed, but the seed of dissension has borne fruit and may still do harm for some time to come—let us hope not for too long.

The Jews of Palestine did not fight for political independence for its own sake, but they fought to achieve free immigration for the Jews of many countries where their very existence was in danger; free immigration also for all those who were longing for a life among their own. It is no exaggeration to say that they fought to make possible a sacrifice perhaps unique in history.

I do not speak of the loss in lives and property fighting an opponent who was numerically far superior, nor do I mean the exhausting toil which is the pioneer's lot in a neglected arid country. I am thinking of the additional sacrifice that a population living under such conditions has to make in order to receive, in the course of eighteen months, an influx of immigrants which comprise more than one third of the total Jewish population of the country. In order to realize what this means you have only to visualize a comparable feat of the American Jews. Let us assume there were no laws limiting the immigration into the United States; imagine that the Jews of this country volunteered to receive more than one million Jews from other countries in the course of one year and a half, to take care of them, and to integrate them into the economy of this country. This would be a tremendous achievement, but still very far from the achievement of our brethren in Israel. For the United States is a big, fertile country, sparsely populated with a high living standard and a highly developed productive capacity, not to compare with small Jewish Palestine whose inhabitants, even without the additional burden of mass immigration, lead a hard and frugal life, still threatened by enemy attacks. Think of the privations and personal sacrifices which this voluntary act of brotherly love means for the Jews of Israel.

The economic means of the Jewish Community in Israel do not suffice to bring this tremendous enterprise to a successful end. For a hundred thousand out of more than three hundred thousand persons who immigrated to Israel since May 1948 no homes or work could be made available. They had to be concentrated in improvised camps under conditions which are a disgrace to all of us.

It must not happen that this magnificent work breaks down because the Jews of this country do not help sufficiently or quickly enough. Here, to my mind, is a precious gift with which all Jews have been presented: the opportunity to take an active part in this wonderful task.

A Biography of Albert Einstein

ALBERT EINSTEIN (1879–1955) is among modern history's greatest and most influential minds. He authored more than 450 scholarly works during his lifetime, and his advancements in science—including the revolutionary Theory of Relativity and E=mc2, which described for the first time the relationship between an object's mass and its energy—have earned him renown as "the father of modern physics."

Born in Ulm, in southwest Germany, Einstein moved to Munich with his family as an infant. As a child, Einstein spoke so infrequently that his parents feared he had a learning disability. But despite difficulties with speech, he was consistently a top student and showed an early aptitude for mathematics and physics, which he later studied at the Swiss Federal Institute of Technology in Zurich after renouncing his German citizenship to avoid military service in 1896.

After graduation, Einstein married his college girlfriend, Mileva Marić, and they had three children. He attended the University of Zurich for his doctorate and worked at the patent office in Bern, a post he left in 1908 for a teaching position at the University of Bern, followed by a number of professorships throughout Europe that ultimately led him back to Germany in 1914. By this time, Einstein had already become recognized throughout the world for his groundbreaking papers on special relativity, the photoelectric effect, and the relationship between energy and matter. He won the Nobel Prize in Physics in 1921.

In 1933, Einstein escaped Nazi Germany and immigrated to the United States with his second wife, Elsa Löwenthal, whom he had married in 1919. He accepted a position at Princeton University in New Jersey, where he stayed for the remainder of his life. At Princeton, Einstein dedicated himself to finding a unified field theory and played a key role in America's development of atomic weapons. He also campaigned for civil rights as a member of the NAACP and was an ardent supporter of Israel's Labor Zionist Movement.

Still, Einstein maintained a special affinity for his homeland. His connection to all things German and, in particular, to the scientific community in Berlin was probably the reason that throughout his years in America he so strongly valued his relationships with other German-speaking immigrants. He maintained a deep friendship with the founder of Philosophical Library, Dr. Dagobert D. Runes, who, like Einstein, was a humanist, a civil rights pioneer, and an admirer of Baruch Spinoza. Consequently, many of Albert Einstein's works were published by Philosophical Library.

At the time of Einstein's death in 1955, he was universally recognized as one of history's most brilliant and important scientists.

*Einstein with friends Marcel Grossmann, Eugen Grossmann,
and Gustav Geissler in the garden of the Grossmann home in
Thalwil, Switzerland, around 1899. Einstein's discussions with
Marcel about elliptic geometry provided one of the sparks that
led to Einstein's development of the General Theory of Relativity.*

Einstein with his first wife, Mileva Marić, and their son Hans Albert, in 1904. Their second son, Eduard, would be born six years later.

A twenty-six-year-old Einstein during the time he was employed at the Bern patent office, in 1905.

Paper silhouettes created by Einstein in 1919, the year of his marriage to his second wife, Elsa. The silhouettes depict, from left to right, himself, Elsa, and his stepdaughters Ilse and Margot.

Einstein lecturing in Vienna, Austria, in January of 1921, the same year he won the Nobel Prize in Physics. 1921 also marked the year of Einstein's first visit to New York City, followed by weeks of lectures at some of the East Coast's most prestigious universities.

Einstein with Elsa in Migdal, Israel, on February 12, 1923.

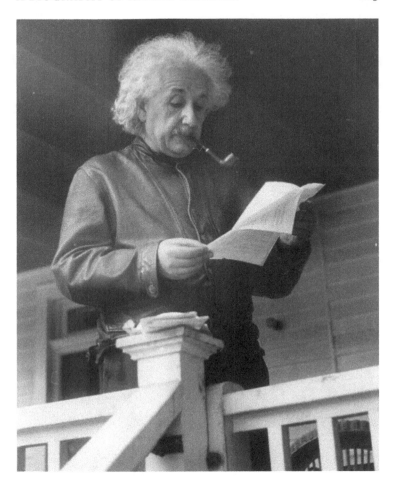

Einstein smoking a pipe on the porch of his home in Princeton, New Jersey, in 1938. He was a very ardent pipe smoker and treasured the ritual of selecting different tobaccos and preparing them to be smoked.

Einstein with his friends poet Itzik Feffer and actor Solomon Mikhoels, in 1943.

Einstein in his Princeton study on the day that he received his honorary degree from the Hebrew University of Jerusalem, in 1949.

Einstein receiving the honorary degree from Israel S. Wechsler while at his Princeton home in 1949.

A portrait of Einstein at the Yeshiva University inauguration dinner for the Albert Einstein College of Medicine, at Princeton Inn on March 15, 1953.

Acknowledgments

1. From *Monthly Review*; New York, May, 1949.
2. From *Pageant*; New York, January, 1946.
3. From *Science*; Washington, D.C., Winter issue, 1935-36. (Translation prepared by Heinz and Ruth Norden.)
4. From a broadcast over ABC to the Rally of Students for Federal World Government; Chicago, May 24, 1946. Hitherto unpublished.
5. From *One World or None*, edited by Katherine Way and Dexter Masters; Whittlesey House, New York, 1946.
6. From the address delivered at Carnegie Hall, New York, upon receiving the One World Award, April, 1948. Hitherto unpublished.
7. From a speech delivered in Albert Hall, London, October, 1933. Hitherto unpublished.
8. From the message to the Peace Congress of Intellectuals in Wroclav. (This message was never delivered, but was released to the press on August 29, 1948.)
9. From *United Nations World*; New York, October, 1947.
10. From *Moscow New Times*, November 26, 1947; and from *Bulletin of the Atomic Scientists*; Chicago, February, 1948.
11. From a statement to the National Wartime Conference, 1945. Hitherto unpublished.
12. From *The Nation*; New York, October 3, 1934.
13. Written in 1936 for a gathering of university teachers which never took place. Hitherto unpublished.

14. From *Atlantic Monthly*; Boston, November, 1945 and November, 1947. As told to Raymond Swing.

15. From an address at the Fifth Nobel Anniversary Dinner at the Hotel Astor, New York, December 10, 1945. Hitherto unpublished.

16. From an address at the second annual dinner given by the Foreign Press Association to the General Assembly and Security Council of the United Nations, at the Waldorf-Astoria Hotel, New York, November 11, 1947. Hitherto unpublished.

17. From an address delivered at the Conference of the Progressive Education Association, November 23, 1934. Hitherto unpublished.

18. From *Policy*; Chicago, November 27, 1934.

19. From *The American Scholar*; New York, Summer, 1947.

20. From an address to the students of the California Institute of Technology, January 22, 1933. Hitherto unpublished.

21. From *The Manchester Guardian*; Manchester, England, Christmas, 1942.

22. Preface to Johannes Kepler's Letters edited by Mrs. David Baumgardt and as yet unpublished.

23. Statement on the occasion of the Curie Memorial Celebration at the Roerich Museum, New York, November 23, 1935.

24. Statement read at the Memorial Services for Max Planck, April, 1948.

25. From *La Pensee*; Paris, February–March, 1947.

26. From *The Scientific Monthly*; Washington, D.C., Vol. LIV, February, 1942.

27. From *Almanak van het Leidsche Studentencorps* published by S. C. Doesburg Verlag, Leiden, Holland; 1934.

28. Statement on the occasion of Mahatma Gandhi's 70th birthday, 1939. Hitherto unpublished.

29. Statement read at the Nobel Foundation Dinner, December 10, 1946. Hitherto unpublished.

30. From *Collier's*; New York, November 26, 1938.

31. From an address over the Columbia Broadcasting System for the United Jewish Appeal, March 22, 1939.

32. Written in 1934. Hitherto unpublished.

33. Unpublished preface to a Black Book. Written in 1945.

34. From a broadcast for the United Jewish Appeal, April 11, 1943. Hitherto unpublished.

35. From an address delivered at the "Third Seder" celebration of the National Labor Committee for Palestine, at the Commodore Hotel, New York, April 17, 1938, and published in New Palestine; Washington, D.C., April 29, 1938.

36. From *Bulletin of the Society of Polish Jews*; New York, 1944.

37. From a statement read at the unveiling of the Memorial for the Battle of the Warsaw Ghetto; Warsaw, April 19, 1948. Hitherto unpublished.

38. From an address to the Jewish Academy of Sciences and Arts; March 22, 1936. Hitherto unpublished.

39. From a statement read at the Maimonides Jubilee Celebration, New York, April, 1935. Hitherto unpublished. (Translation prepared by Heinz and Ruth Norden.)

40. From *Opinion*; New York, March, 1949.

41. Statement to the Hebrew University in Jerusalem, Israel, on March 15, 1949. Hitherto unpublished.

42. From a letter to the Committee on Unity for Palestine, New York, 1945.

43. From a broadcast for the United Jewish Appeal, over the National Broadcasting Company, November 27, 1949.

PHILOSOPHICAL
LIBRARY

*Philosophical Library's mission is to reintroduce readers to **books of lasting value** by the intellectual icons of the twentieth century, including Albert Einstein, Jean-Paul Sartre, Kahlil Gibran, and André Gide.*

OPEN ROAD
INTEGRATED MEDIA

Open Road Integrated Media is a digital publisher and multimedia content company. Open Road creates connections between authors and their audiences by marketing its ebooks through a new proprietary online platform, which uses premium video content and social media.